AN INTRODUCTION TO NON-ABELIAN CLASS FIELD THEORY

Automorphic Forms of Weight 1 and 2-Dimensional Galois Representations

Series on Number Theory and Its Applications* ISSN 1793-3161

Published

Vol. 7 Geometry and Analysis of Automorphic Forms of Several Variables
Proceedings of the International Symposium in Honor of Takayuki Oda on the Occasion of His 60th Birthday
edited by Yoshinori Hamahata, Takashi Ichikawa, Atsushi Murase & Takashi Sugano

Vol. 8 Number Theory: Arithmetic in Shangri-La
Proceedings of the 6th China–Japan Seminar
edited by S. Kanemitsu, H.-Z. Li & J.-Y. Liu

Vol. 9 Neurons: A Mathematical Ignition
by Masayoshi Hata

Vol. 10 Contributions to the Theory of Zeta-Functions: The Modular Relation Supremacy
by S. Kanemitsu & H. Tsukada

Vol. 11 Number Theory: Plowing and Starring Through High Wave Forms
Proceedings of the 7th China–Japan Seminar
edited by Masanobu Kaneko, Shigeru Kanemitsu & Jianya Liu

Vol. 12 Multiple Zeta Functions, Multiple Polylogarithms and Their Special Values
by Jianqiang Zhao

Vol. 13 An Introduction to Non-Abelian Class Field Theory: Automorphic Forms of Weight 1 and 2-Dimensional Galois Representations
by Toyokazu Hiramatsu & Seiken Saito

*For the complete list of the published titles in this series, please visit:
www.worldscientific.com/series/sntia

Series on Number Theory and Its Applications Vol. 13

AN INTRODUCTION TO NON-ABELIAN CLASS FIELD THEORY

Automorphic Forms of Weight 1 and 2-Dimensional Galois Representations

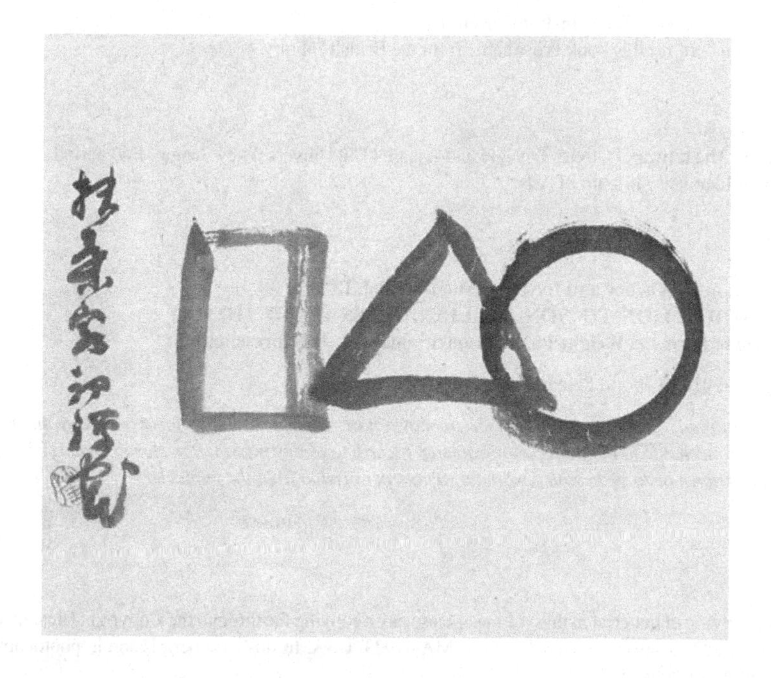

Toyokazu Hiramatsu
Hosei University, Japan

Seiken Saito
Waseda University, Japan

NEW JERSEY • LONDON • SINGAPORE • BEIJING • SHANGHAI • HONG KONG • TAIPEI • CHENNAI • TOKYO

Published by

World Scientific Publishing Co. Pte. Ltd.
5 Toh Tuck Link, Singapore 596224
USA office: 27 Warren Street, Suite 401-402, Hackensack, NJ 07601
UK office: 57 Shelton Street, Covent Garden, London WC2H 9HE

British Library Cataloguing-in-Publication Data
A catalogue record for this book is available from the British Library.

Book cover: The artwork "Circle, Triangle and Square (The Universe)" by Sengai, Edo period.
Courtesy of Idemitsu Museum of Arts.

Series on Number Theory and Its Applications — Vol. 13
AN INTRODUCTION TO NON-ABELIAN CLASS FIELD THEORY
Automorphic Forms of Weight 1 and 2-Dimensional Galois Representations

ISBN 978-981-3142-26-8

Printed in Singapore

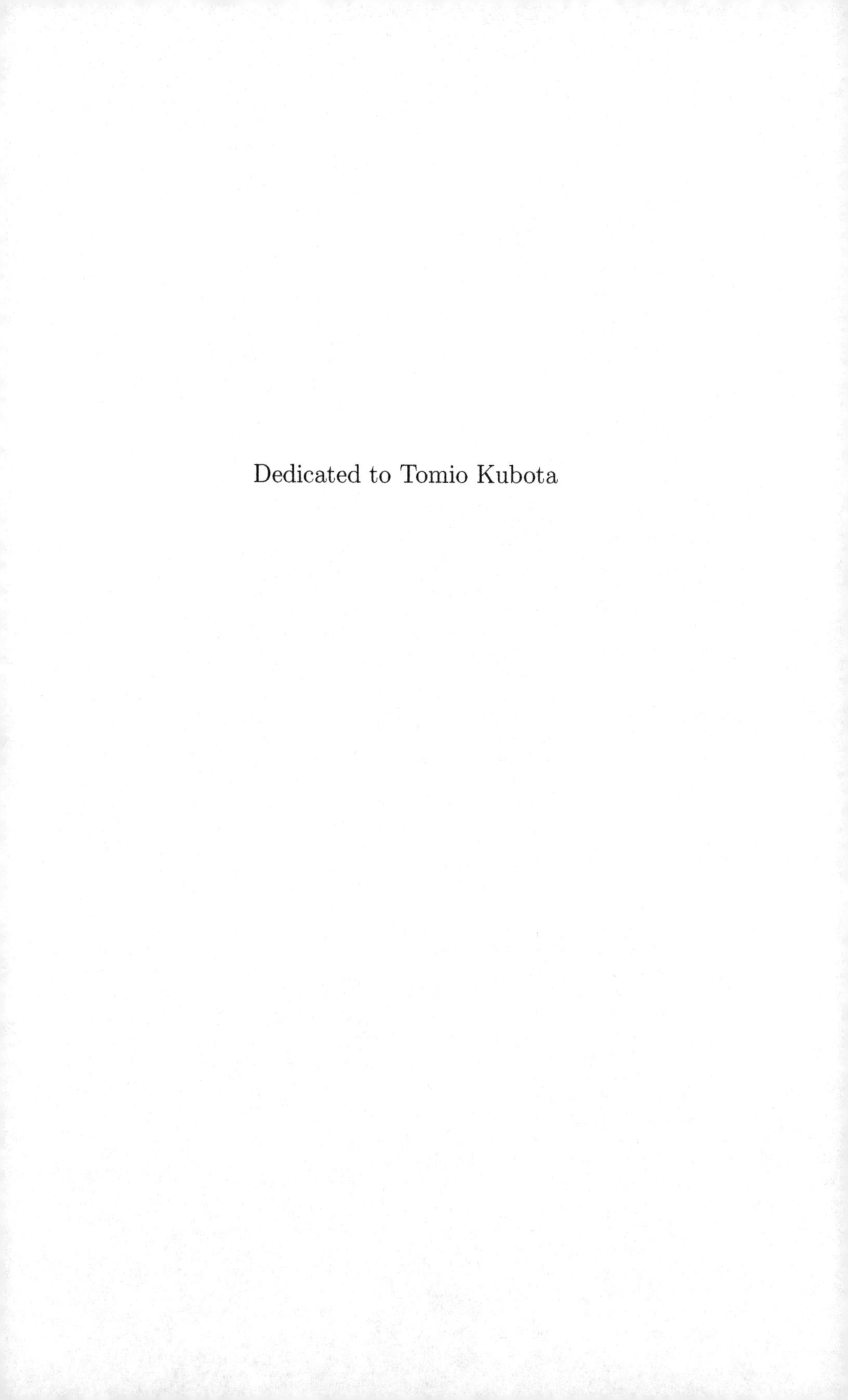
Dedicated to Tomio Kubota

Preface

This monograph is intended to provide a brief exposition of the theory of automorphic forms of weight 1 and their applications in arithmetic. One of the outstanding problems in arithmetic is to generalize class field theory to non-abelian Galois extensions of number fields. This problem was already proposed by Takagi in his talk at the Strasbourg Congress, 1920. Significant progress has been made in recent years by Langlands and others. In the monograph, we discuss some of the relations between this problem and cusp forms of weight 1.

The monograph consists of nine chapters and an appendix. In Chapters 1 and 2, we discuss higher reciprocity laws and arithmetic congruence relations for (non-abelian) dihedral polynomials. In addition, Chapter 1 contains an overview of modular forms and Hecke operators. Chapter 3 will be devoted to the study of Hecke's indefinite modular forms of weight 1, and a relation between positive definite theta series and indefinite theta series.

Let Γ be a fuchsian group of the first kind and let d_1 be the dimension of the linear space of cusp forms of weight 1 on the group Γ. It is not effective to compute the number d_1 by means of the Riemann-Roch theorem. In the final chapter of Part I, we give some formula for d_1 by using the Selberg trace formula, and also discuss $d_1 \bmod 2$.

The first chapter of Part II contains a very remarkable account of various aspects of the theory including Galois representations of odd type, the Artin conjecture, the Langlands program, base change and icosahedral representations. In Chapter 6, we discuss some relations between Maass cusp forms and Galois representations of even type. We also introduce some aspects of automorphic hyperfunctions of weight 1 related to Hecke's theta series. Chapter 7 begins with reviewing some basic facts concerning five conjectures in arithmetic, and we discuss some relations between these conjectures. In Chapter 8, we describe a family of modular series associated with indefinite binary quadratic forms. These series introduced by Polishchuk generate the same space of weight 1 modular forms as Hecke's indefinite theta series.

The dimension of the space of Hilbert modular type cusp forms has been calculated in most of cases, but not yet for the case of weight 1. In Chapter 9, we shall discuss the dimension for this remaining case. Let K be a real quadratic field and $\mathfrak{O}_K$ be the ring of integers in K. By the technical reason, we assume that the class number of K is equal to 1. The purpose of this chapter is more precisely to study the dimension for the Hilbert modular type cusp forms of weight 1 with respect to the Hilbert modular group $\mathrm{SL}_2(\mathfrak{O}_K)$ through Selberg's trace formula.

The articles in the Appendix are the reproduction of a manuscript related to the first author's Göttingen talk.

Kobe *Toyokazu Hiramatsu*
Tokyo *Seiken Saito*
June, 2015

Contents

Part I

Chapter 1

Higher reciprocity laws

Let $f(x)$ be a monic irreducible polynomial with integer coefficients and let p be a prime number. Reducing the coefficients of $f(x)$ modulo p, we obtain a polynomial $f_p(x)$ with coefficients in the p-element field $\mathbb{F}_p$. We define $\mathrm{Spl}\{f(x)\}$ to be the set of all primes such that the polynomial $f_p(x)$ factors into a product of distinct linear polynomials over the field $\mathbb{F}_p$. What is the rule to determine the primes belonging to $\mathrm{Spl}\{f(x)\}$? We may call its answer a higher reciprocity law for the polynomial $f(x)$. For example, the usual law of quadratic reciprocity in the elementary number theory gives a 'reciprocity law' in the above sense: Let q be an odd prime. Then the set $\mathrm{Spl}\{x^2 - q\}$ can be described by congruence conditions modulo q if $q \equiv 1 \pmod 4$ and modulo $4q$ if $q \equiv 3 \pmod 4$. The polynomial $f(x)$ is called an abelian polynomial if its Galois group is abelian. Then, the next theorem, a natural consequence from class field theory over the rational number field $\mathbb{Q}$, is known:

Theorem. *The set $\mathrm{Spl}\{f(x)\}$ can be described by congruence relations for modulus depending only on $f(x)$ if and only if $f(x)$ is abelian.*

If $f(x)$ is a polynomial with non-abelian Galois group, then very little can be said about the set $\mathrm{Spl}\{f(x)\}$. We may call any rule to determine the set $\mathrm{Spl}\{f(x)\}$ a *higher reciprocity law* for non-abelian polynomial $f(x)$. The main purpose of this chapter is to give some examples of higher reciprocity law for non-abelian polynomials arising from the dihedral cusp forms of weight 1.

1.1 Some examples of non-abelian case

1.1.1 $f(x) = x^3 - d$

Example 1.1 (Spl$\{x^3-2\}$). Let $\omega = (-1+\sqrt{-3})/2$ and consider the ring $\mathbb{Z}[\omega] = \{a + b\omega \mid a, b \in \mathbb{Z}\}$. Let π be a prime in $\mathbb{Z}[\omega]$. If $N(\pi) \neq 3$, the cubic residue of α modulo π is given by

(i) $(\alpha/\pi)_3 = 0$, if $\pi \mid \alpha$,

(ii) $\alpha^{(N(\pi)-1)/3} \equiv (\alpha/\pi)_3 \pmod{\pi}$, with $(\alpha/\pi)_3$ equal to 1, ω or ω^2.

A prime π is called primary if $\pi \equiv 2 \pmod 3$. Then we can state

Theorem (Cubic Reciprocity Law). *Let π_1 and π_2 be primary, $N(\pi_1)$, $N(\pi_2) \neq 3$, and $N(\pi_1) \neq N(\pi_2)$. Then*

$$\left(\frac{\pi_2}{\pi_1}\right)_3 = \left(\frac{\pi_1}{\pi_2}\right)_3.$$

Now we have the following by the above cubic reciprocity law:

Theorem 1.1. Spl$\{x^3-2\}$

$$\begin{aligned} &= \{p \mid p \equiv 1 \pmod 3,\ p = x^2 + 27y^2\ x, y \in \mathbb{Z}\} \\ &= \{p \mid p \equiv 1 \pmod 3,\ \left(\tfrac{2}{\pi}\right)_3 = 1 \textit{ for } p = \pi\bar{\pi}\} \\ &= \{p \mid p \equiv 1 \pmod 3,\ a(p) = 2\}, \end{aligned}$$

where $a(p)$ denotes the p-th coefficient of the expansion

$$\eta(6\tau)\eta(18\tau) = \sum_{n=1}^{\infty} a(n)q^n, \quad q = e^{2\pi i\tau}$$

with the Dedekind eta function $\eta(\tau)$ defined by

$$\eta(\tau) = q^{1/24} \prod_{n=1}^{\infty} (1 - q^n), \quad \operatorname{Im}\tau > 0.$$

Proof. The first half. Let p be a rational prime such that $p \equiv 1 \pmod 3$. Then $p = \pi\bar{\pi}$ in $\mathbb{Z}[\omega]$. Suppose that π is primary. Then, by the law of cubic reciprocity, we have the following two facts:

(1) $x^3 \equiv 2 \pmod{\pi}$ is solvable if and only if $\pi \equiv 1 \pmod 2$;

(2) If $p \equiv 1 \pmod 3$, then $x^3 \equiv 2 \pmod p$ is solvable if and only if there are integers x and y such that $p = x^2 + 27y^2$.

By (1) and (2), we have the first half of Theorem 1.1

The latter half. By the *Euler pentagonal number theorem*, we have

$$\eta(6\tau)\eta(18\tau) = \sum_{m,n\in\mathbb{Z}} (-1)^{m+n} q^{\{(6m+1)^2+3(6n+1)^2\}/4}.$$

Let denote by $A(p)$ the number of solutions (m, n) of

$$(6m+1)^2 + 3(6n+1)^2 = 4p.$$

Then we have easily the following assertions:

(i) $A(p) = 2$ and $m + n$ is even if $p = x^2 + 27y^2$;

(ii) $A(p) = 1$ and $m + n$ is odd if $p \neq x^2 + 27y^2$.

Therefore we have the latter half of Theorem 1.1. □

Example 1.2 (Cubic residuacity). Let d be a non-cubic integer and put $K = k(\sqrt[3]{d})$ for $k = \mathbb{Q}(\sqrt{-3})$. Then K is a splitting field of $f(x) = x^3 - d$ over $\mathbb{Q}$ with the Galois group $\mathrm{Gal}(K/\mathbb{Q}) \cong S_3$, the symmetric group of order 3, and K/k is a cyclic extension of degree 3. Hence K is the class field over k with conductor $\mathfrak{f} = (3d)$. We denote by $T_{\mathfrak{f}}$ the ideal group corresponding to K.

For any odd prime p except the divisors of $\mathfrak{f}$, we know that f mod p can factor over the p-element field $\mathbb{F}_p$ in one of the three ways:

(i) (Linear)(Quadratic) if $p \equiv 2 \pmod 3$,

(ii) Three linear factors if $p \equiv 1 \pmod 3$ and $\left(\dfrac{d}{p}\right)_3 = 1$,

(iii) Irreducible otherwise.

If $p \equiv 1 \pmod 3$, then p splits in k as $p = \mathfrak{p}_p\bar{\mathfrak{p}}_p$, and we obtain

$$\begin{aligned}\mathfrak{p}_p \in T_{\mathfrak{f}} &\iff \mathfrak{p}_p \text{ splits completely in } K\\ &\iff f(x) \text{ has exactly 3 linear factors mod } p\\ &\iff f(x) \equiv 0 \pmod p \text{ has an integral solution}\\ &\iff \left(\frac{d}{p}\right)_3 = 1.\end{aligned}$$

Now we put

$$\begin{aligned}I_{\mathfrak{f}} &= \{(\alpha) \mid (\alpha, \mathfrak{f}) = 1\},\\ J_{\mathfrak{f}} &= \{(\alpha) \in I_{\mathfrak{f}} \mid \alpha \equiv a \pmod{\mathfrak{f}} \text{ for some } a \in \mathbb{Z}\},\\ P_{\mathfrak{f}} &= \{(\alpha) \in I_{\mathfrak{f}} \mid \alpha \equiv 1 \pmod{\mathfrak{f}}\}.\end{aligned}$$

Then we have the following table:

field	corresponding ideal group	index
maximal ray class field	$P_{\mathfrak{f}}$	$d+\left(\frac{d}{3}\right)$
ring class field	$J_{\mathfrak{f}}$	$\frac{1}{3}\left(d-\left(\frac{d}{3}\right)\right)$
K	$T_{\mathfrak{f}}$	3
k	$I_{\mathfrak{f}}$	

Hence we observe the group $T_{\mathfrak{f}}$ as the union of $\dfrac{1}{3}\left(d-\left(\dfrac{d}{3}\right)\right)$ cosets of $J_{\mathfrak{f}}$. And if d is prime then it follows that $I_{\mathfrak{f}}/P_{\mathfrak{f}}$ is the direct product of two cyclic groups or a cyclic group according to whether $d \equiv 1 \pmod 3$ or not.

Let χ be an ideal character of $I_{\mathfrak{f}}/T_{\mathfrak{f}}$, and put

$$L(s,\chi) = \sum_{\mathfrak{a}} \chi(\mathfrak{a}) N_{K/\mathbb{Q}}(\mathfrak{a})^{-s} = \sum_{n=1}^{\infty} a_n n^{-s},$$

where $\mathfrak{a}$ runs all integral ideals in $I_{\mathfrak{f}}$. Since $L(s,\chi)$ has an Euler product expansion $(p \nmid \mathfrak{f})$

$$L(s,\chi) = \prod_{p\equiv 2\,(3)} (1-p^{-2s})^{-1} \prod_{\substack{p\equiv 1\,(3)\\ \left(\frac{d}{3}\right)_3=1}} (1-p^{-s})^{-2} \prod_{\substack{p\equiv 1\,(3)\\ \left(\frac{d}{3}\right)_3\neq 1}} (1+p^{-s}+p^{-2s})^{-1},$$

we have

$$a_p = 0 \quad \text{if } p \equiv 2 \pmod 3,$$

$$a_p = 2 \quad \text{if } p \equiv 1 \pmod 3 \text{ and } \left(\frac{d}{p}\right)_3 = 1,$$

$$a_p = -1 \text{ if } p \equiv 1 \pmod 3 \text{ and } \left(\frac{d}{p}\right)_3 \neq 1.$$

Therefore,

$$\#\{\alpha \in \mathbb{F}_p \mid f(\alpha) = 0\} = a_p + 1.$$

After this we shall denote the number of elements of a finite set S by $\#S$. Put $q = e^{2\pi i\tau}$ for $\mathrm{Im}(\tau) > 0$. Then the corresponding form

$$\theta(\tau) = \sum_{\mathfrak{a}} \chi(\mathfrak{a}) q^{N(\mathfrak{a})} = \sum_{n=1}^{\infty} a_n q^n$$

of $L(s,\chi)$ is a cusp form of weight 1 and character $\left(\frac{-3}{*}\right)$ for the congruence subgroup $\Gamma_0(3^3d^2)$ (for definitions, cf. Section 1.2). Hence we can obtain that the cubic residuacity of d is determined by the reduction modulo 2 of the Fourier coefficients of the above θ. Then we set

Problem. *Express $\theta(\tau)$ explicitly by using the known functions and consider the cubic residuacity more concretely.*

Example 1.3 (Spl$\{x^3-2\}$ (Revisited)). In this case $d=2$, it follows that $T_{\mathfrak{f}} = J_{\mathfrak{f}} = P_{\mathfrak{f}}$, and we have

$$I_{\mathfrak{f}}/P_{\mathfrak{f}} = \langle \mathfrak{p}_7 P_{\mathfrak{f}} \rangle,$$

where $\mathfrak{p}_7 = (2+\sqrt{-3})$. By a simple calculation, we see that

$$a+b\omega \in P_{\mathfrak{f}} \Longleftrightarrow a \equiv 3 \text{ and } b \equiv 1 \pmod 6$$

and

$$a+b\omega \in \mathfrak{p}_7 P_{\mathfrak{f}} \Longleftrightarrow a \equiv 0 \text{ and } b \equiv 1 \pmod 6,$$

where $\omega = (1+\sqrt{-3})/2$. Thus we can exchange a and b for $3a$ and $6b+1$ respectively. And since

$$N(3a+(6b+1)\omega) = \{(6(a+b)+1)^2 + 3(6b+1)^2\}/4,$$

we obtain that

$$\begin{aligned}
\theta(\tau) &= \sum_{\mathfrak{a}\in P_{\mathfrak{f}}} q^{N(\mathfrak{a})} - \sum_{\mathfrak{a}\in \mathfrak{p}_7 P_{\mathfrak{f}}} q^{N(\mathfrak{a})} \\
&= \sum_{a,b\in\mathbb{Z}} (-1)^a q^{N(3a+(6b+1)\omega)} \\
&= \sum_{a,b\in\mathbb{Z}} (-1)^{a+b} q^{((6a+1)^2+3(6b+1)^2)/4} \\
&= \eta(6\tau)\eta(18\tau).
\end{aligned}$$

Example 1.4 (Spl$\{x^3-3\}$). In this case $d=3$, $T_{\mathfrak{f}} = J_{\mathfrak{f}}$, and we have

$$I_{\mathfrak{f}}/J_{\mathfrak{f}} = \langle \mathfrak{p}_7 J_{\mathfrak{f}} \rangle.$$

For an integral ideal $\mathfrak{a}$ belonging to $I_{\mathfrak{f}}$, we set

$$\mathfrak{a} = (a) \text{ and } \alpha = (x+3y\sqrt{-3})/2$$

where $x \equiv 2 \pmod 3$ and $x \equiv y \pmod 2$.
Then, by an easy calculation, we see that

$$\begin{aligned}(\alpha) \in J_f &\iff y \equiv 0 \pmod 3,\\ (\alpha) \in \mathfrak{p}_7 J_f &\iff y \equiv 1 \pmod 3,\\ (\alpha) \in \mathfrak{p}_7^2 J_f &\iff y \equiv 2 \pmod 3.\end{aligned}$$

Hence we obtain

$$\chi(\mathfrak{a}) = \zeta^y$$

for $\zeta = e^{2\pi i/3}$, and hence

$$\begin{aligned}\theta(\tau) &= \sum_{\substack{x\equiv 2(3)\\ x\equiv y(2)}} \zeta^y q^{(x^2+27y^2)/4}\\ &= \frac{1}{2}\{\sum_{\substack{x\equiv 2(3)\\ y}} \zeta^y q^{(x^2+27y^2)/4} + \sum_{\substack{x\equiv 2(3)\\ y}} (-1)^{x+y}\zeta^y q^{(x^2+27y^2)/4}\}\\ &= \frac{1}{2}\{\sum_{x\equiv 2(3)} q^{x^2/4}\cdot\sum_{y}\zeta^y q^{27y^2/4}\\ &\quad + \sum_{x\equiv 2(3)} (-1)^x q^{x^2/4}\cdot\sum_{y}(-1)^y\zeta^y q^{27y^2/4}\}\\ &= \frac{1}{8}\{(\theta_3(\tau/2)-\theta_3(9\tau/2))\,(3\theta_3(243\tau/2)-\theta_3(27\tau/2))\\ &\quad + (\theta_0(\tau/2)-\theta_0(9\tau/2))\,(3\theta_0(243\tau/2)-\theta_0(27\tau/2))\}\\ &= \frac{1}{4}\{(\theta_3(2\tau)-\theta_3(18\tau))\,(3\theta_3(486\tau)-\theta_3(54\tau))\\ &\quad + (\theta_2(2\tau)-\theta_2(18\tau))\,(3\theta_2(486\tau)-\theta_2(54\tau))\},\end{aligned}$$

where

$$\theta_0(\tau) = \sum_{m\in\mathbb{Z}}(-1)^m q^{m^2/2}, \quad \theta_3(\tau) = \sum_{m\in\mathbb{Z}} q^{m^2/2} \text{ and}$$

$$\theta_2(\tau) = \sum_{m\in\mathbb{Z}} q^{(m+1/2)^2/2}.$$

1.1.2 $f(x) = 4x^3 - 4x^2 + 1$

E2.1 We put

$$\eta(\tau)^2\eta(11\tau)^2 = \sum_{n=1}^{\infty} b(n)q^n, \quad q = e^{2\pi i\tau}.$$

By the Euler pentagonal number theorem, we have

$$\sum_{n=1}^{\infty} b(n)q^n \equiv \sum_{u,v\in\mathbb{Z}} q^{\{(6u+1)^2+11(6v+1)^2\}/12} \pmod 2.$$

Let $B(n)$ be the number of solutions (u, v) of

$$(6u+1)^2 + 11(6v+1)^2 = 12n.$$

When n is prime $p \equiv 2, 6, 7, 8, 10 \pmod{11}$, we see that $B(p) = 0$. For the remaining cases, we have the following

Lemma. *Let p be a prime such that $p \equiv 1, 3, 5, 9 \pmod{11}$. Then either $p \equiv x^2+11y^2$ or $p \equiv 3u^2+2uv+4v^2$, and two cases are mutually exclusive, namely, either p or $3p$ is of the form $x^2 + 11y^2$ for some integers x and y. Moreover, the following assertions hold:*

(i) *$B(p) = 2$ and $u + v$ is even if $p = x^2 + 11y^2$;*

(ii) *$B(p) = 1$ and $u + v$ is odd if $3p = X^2 + 11Y^2$.*

Proof. The first half. Since $(-11/p) = 1$, we have

$$p = a^2 + ab + 3b^2$$

for some integers a and b. If b is even, then

$$\begin{aligned} p &= \left(a + \frac{b}{2}\right)^2 + 11\left(\frac{b}{2}\right)^2 \\ &= x^2 + 11y^2 \quad (x, y \in \mathbb{Z}). \end{aligned}$$

For b odd,

$$3p = \left(3b + \frac{a}{2}\right)^2 + 11\left(\frac{a}{2}\right)^2 \quad (a\text{: even})$$

or

$$3p = \left(3b - \frac{a+b}{2}\right)^2 + 11\left(\frac{a+b}{2}\right)^2 \quad (a\text{: odd}),$$

and hence $3p = X^2 + 11Y^2$ for some integers X and Y. Since matrices $\begin{pmatrix} 1 & 0 \\ 0 & 11 \end{pmatrix}$ and $\begin{pmatrix} 3 & 1 \\ 1 & 4 \end{pmatrix}$ are not equivalent, the two cases are mutually exclusive.

The latter half. Put

$$D(p) = \{(s,t) \mid s^2 + 11t^2 = 4p,\ s + t \equiv 2 \pmod{12}\}.$$

Then we see at once that $B(p) = \#D(p)$. If $p = x^2 + 11y^2$, then there are four solutions of the equation $s^2 + 11t^2 = 4p$. Moreover, $s + t \equiv 2 \pmod 4$ and $s + t \not\equiv 0 \pmod 3$. Therefore $\#D(p) = 2$. If $3p = X^2 + 11Y^2$, then $X \equiv Y \pmod 3$, $X \not\equiv Y \pmod 2$ and

$$4p = \left(\frac{X + 11Y}{3}\right)^2 + 11\left(\frac{X - Y}{3}\right)^2.$$

Hence there is the only solution of $s^2+11t^2 = 4p$ such that $s+t \equiv 2 \pmod 4$ and $s + t \equiv 2 \pmod 3$. Therefore $\#D(p) = 1$. Hence we have

$$B(p) = \begin{cases} 2, \text{ if } p = x^2 + 11y^2, \\ 1, \text{ if } 3p = X^2 + 11Y^2. \end{cases}$$

Next it is obvious that

$$\begin{aligned} p &= (3u^2 + u) + 11(3v^2 + v) + 1 \\ &= \left(\frac{u + 11v}{2} + 1\right)^2 + 11\left(\frac{v - u}{2}\right)^2. \end{aligned}$$

Therefore, if $u + v$ is even then

$$p = x^2 + 11y^2 \quad (x, y \in \mathbb{Z}).$$

On the other hand,

$$\begin{aligned} 3p &= 3(3u^2 + u) + 33(3v^2 + v) + 3 \\ &= \left(\frac{-5u + 11v + 11}{2}\right)^2 + 11\left(\frac{u + 5v + 1}{2}\right)^2. \end{aligned}$$

Therefore, if $u + v$ is odd then $3p = X^2 + 11Y^2$ $(X, Y \in \mathbb{Z})$. □

Let E be the elliptic curve over $\mathbb{Q}$ defined by

$$y^2 = f(x), \quad f(x) = 4(x^3 - x^2) + 1,$$

which is derived from Tate's form $y^2 + y = x^3 - x^2$. Let p be a good prime for E and $\tilde{E}_p$ denote the reduction modulo p of E which is an elliptic curve over $\mathbb{F}_p$. It is a special (proved) case of the Taniyama-Shimura conjecture that the number N_p of $\mathbb{F}_p$-rational points of $\tilde{E}_p$ is given by

$$N_p = p - b(p).$$

Then it is clear that (1) N_p is even if $f(x)$ is irreducible $(\bmod\, p)$, (2) N_p is odd if $f(x)$ has exactly one or three linear factors $(\bmod\, p)$. Therefore we have the following

Theorem 1.2. *Let p be any odd prime, except 11 and put $f_p(x) = f(x) \bmod p$. Then $f_p(x)$ can factor over $\mathbb{F}_p$ in one of the three ways:*

(i) *exactly one linear factor if* $\left(\frac{-11}{p}\right) = -1$;

(ii) *exactly 3 linear factors if* $\left(\frac{-11}{p}\right) = 1$ *and* $p = x^2 + 11y^2$ $(x, y \in \mathbb{Z})$;

(iii) *no linear factor if* $\left(\frac{-11}{p}\right) = 1$ *and* $3p = X^2 + 11Y^2$ $(X, Y \in \mathbb{Z})$.

Corollary. $\mathrm{Spl}\{4x^3 - 4x^2 + 1\}$

$$= \left\{ p \,\middle|\, \left(\frac{-11}{p}\right) = 1,\ p = x^2 + 11y^2 \right\}$$

$$= \left\{ p \,\middle|\, \left(\frac{-11}{p}\right) = 1,\ b(p) \equiv 0 \pmod 2 \right\}.$$

E2.2 We start with

$$\begin{aligned} \eta(2\tau)\eta(22\tau) &= q \prod_{n=1}^{\infty} (1 - q^{2n})(1 - q^{22n}) \\ &= q \sum_{u,v \in \mathbb{Z}} (-1)^{u+v} q^{(3u^2+u)+11(3v^2+v)} \\ &= \sum_{u,v \in \mathbb{Z}} (-1)^{u+v} q^{\{(6u+1)^2+11(6v+1)^2\}/12} \\ &= \sum_{n=1}^{\infty} c(n) q^n, \end{aligned}$$

where $q = e^{2\pi i \tau}$. Then by Lemma, it is immediate that

$$c(p) = \begin{cases} 0, \text{ if } \left(\frac{-11}{p}\right) = -1, \\ 2, \text{ if } \left(\frac{-11}{p}\right) = 1 \text{ and } p = x^2 + 11y^2 \ (x, y \in \mathbb{Z}), \\ -1, \text{ if } \left(\frac{-11}{p}\right) = 1 \text{ and } 3p = X^2 + 11Y^2 \ (X, Y \in \mathbb{Z}). \end{cases}$$

We can now state

Theorem 1.3 ([40]). *Let p be any odd prime, except 11. Then we have the following arithmetic congruence relation*

$$\#\{x \in \mathbb{F}_p | 4x^3 - 4x^2 + 1 = 0\} = c(p)^2 - \left(\frac{-11}{p}\right) = c(p) + 1.$$

Proof. In place of $f(x) = 4x^3 - 4x^2 + 1$, we shall consider

$$h(x) = -2f\left(\frac{1-x}{2}\right) = x^3 - x^2 - x - 1.$$

The polynomial $h(x)$ has discriminant -44. Let $h_p(x)$ be a reduction modulo p of $h(x)$ and let K_h be a splitting field of $h_p(x)$ over the field $\mathbb{F}_p$. Then it can easily be seen that

$$\left(\frac{-11}{p}\right) = -1 \Longleftrightarrow [K_h : \mathbb{F}_p] = 2$$

$$\Longleftrightarrow h_p(x) \text{ has exactly one linear factor over } \mathbb{F}_p.$$

Next we consider the case of $\left(\frac{-11}{p}\right) = 1$. Let L_h be a splitting field of $h(x)$ over $\mathbb{Q}$. Put $k = \mathbb{Q}(\sqrt{-11})$, and observe that L_h is an abelian extension over k of degree 3. Considering L_h as a class field of k, we denote by H its corresponding class group and by f a conductor of H. Since 2 is only ramified in L_h, we thus obtain $f = (2)$. Hence

$$H = \{(\alpha) : \text{ideals in } k \mid \alpha \equiv 1 \,(\mathrm{mod}\, 2)\}.$$

By the assumption $\left(\frac{-11}{p}\right) = 1$, we also have

$$p = \mathfrak{p}\bar{\mathfrak{p}} \text{ in } k,$$

where $\mathfrak{p}$ denotes a prime ideal in k and $\bar{\mathfrak{p}}$ a conjugate of $\mathfrak{p}$; and moreover

$$p \in H \Longleftrightarrow \mathfrak{p} \text{ splits completely in } L_h.$$

Now we put $\mathfrak{p} = (\pi)$ with $\pi = a + b\omega$, where $\omega = (-1+\sqrt{-11})/2$, a and b are rational integers. Then we see from the above result that

$$\begin{aligned}
\mathfrak{p} \in H &\Longleftrightarrow \pi \equiv 1 \,(\mathrm{mod}\, 2) \\
&\Longleftrightarrow b \equiv 0 \,(\mathrm{mod}\, 2) \\
&\Longleftrightarrow p = N(\pi) = x^2 + 11y^2 \; (x, y \in \mathbb{Z}) \\
&\Longleftrightarrow \mathfrak{p} \text{ splits completely in } L_h \\
&\Longleftrightarrow h(x) \text{ has exactly 3 linear factors } (\mathrm{mod}\, \mathfrak{p}) \\
&\Longleftrightarrow h_p(x) \text{ has exactly 3 linear factors over } \mathbb{F}_p.
\end{aligned}$$

Finally, we suppose b is an odd integer in the expression $p = N(\pi) = a^2 + ab + 3b^2$. Then, $3p = X^2 + 11Y^2$ for some integers X and Y, and hence

$$3p = X^2 + 11Y^2 \Longleftrightarrow h_p(x) \text{ has no linear factor over } \mathbb{F}_p.$$

□

Corollary. $\mathrm{Spl}\{4x^3 - 4x^2 + 1\} = \{p \mid c(p) = 2,\, p \neq 2, 11\}$.

Remark 1.1. Let

$$f(x) = x^3 + ax^2 + bx + c \quad (a, b, c \in \mathbb{Z})$$

be an irreducible polynomial whose splitting field K_f is Galois extension over $\mathbb{Q}$ with $\mathrm{Gal}\,(K_f/\mathbb{Q}) \cong S_3$ and contains an imaginary quadratic field k. Let $L(s, \rho)$ be the L-function associated with the representation

$$\rho : \mathrm{Gal}\,(K_f/\mathbb{Q}) \longrightarrow \mathrm{GL}_2(\mathbb{C})$$

with conductor N. Then there exists normalized newform $F(z)$ on $\Gamma_0(N)$ of weight 1 and character $\det \rho$. Now, bringing two objects **E2.1** and **E2.2** into unity, Koike obtained the following arithmetic congruence relation for $f(x)$ ([62]):

Theorem 1.4. *Let M be the product of all primes which appear in a, b and c and let p be any prime such that $p \nmid MN$. Then we have*

$$\#\{\alpha \in \mathbb{F}_p | f(\alpha) = 0\} = a(p)^2 - \left(\frac{-D}{p}\right),$$

where $-D$ denotes the discriminant of k and $a(p)$ denotes the p-th Fourier coefficient of $F(z)$:

$$F(z) = \sum_{n=1}^{\infty} a(n) e^{2\pi i n z}.$$

Corollary. *Let p be any prime such that $p \nmid MN$. Then*

$$\mathrm{Spl}\{f(x)\} = \{p : \mathit{prime} | a(p) = 2\}$$

up to finite set of primes.

1.1.3 $f(x) = x^4 - 2x^2 + 2$

First we recall some known results which appeared in Smith's Number Theory Report.[1]

(i)

$$\begin{aligned} \eta(8\tau)\eta(16\tau) &= \sum_{a,b\in\mathbb{Z}} (-1)^b q^{(4a+1)^2+8b^2} \\ &= \sum_{\alpha,\beta\in\mathbb{Z}} (-1)^{\alpha+\beta} q^{(4\alpha+1)^2+16\beta^2}, \end{aligned}$$

where $q = e^{2\pi i\tau}$;

[1] H. J. S. Smith: Report on the theory of numbers VI, Reports of the British Association for 1865, pp. 322-375, §128: Theorems of Jacobi on Simultaneous Quadratic Forms.

(ii) Let $d(n)$ be the n-th Fourier coefficient of $\eta(8)\eta(16\tau)$ at ∞. Then $d(n)$ is multiplicative and has the following properties:

(1) $d(p) = 2\varepsilon(-1)^{(p-1)/8}$ if $p \equiv 1 \pmod 8$, here $\varepsilon \equiv 2^{(p-1)/4} \pmod p$;
(2) $d(p^{2v}) = (-1)^v$ if $p \equiv 3 \pmod 8$;
(3) $d(p^{2v}) = 1$ if $p \equiv 5, 7 \pmod 8$.

The results (ii) is the first instance of an explicit computation of the Fourier coefficients of a cusp form of weight 1 which is of interest from the point of view of history. Let k be an imaginary quadratic field, say $k = \mathbb{Q}(\sqrt{-p})$ with a prime number $p \equiv 1 \pmod 8$, and let h be the class number of k. We put

$$p = (4a+1)^2 + 8b^2 = (4\alpha+1)^2 + 16\beta^2.$$

Then it is easy to see that

$$\begin{aligned} b \equiv 0 \pmod 2 &\iff \alpha + \beta \equiv 0 \pmod 2 \\ &\iff \left(\frac{-4}{p}\right)_8 = 1 \\ &\iff h \equiv 0 \pmod 8, \end{aligned}$$

where $\left(\frac{-}{p}\right)_8$ denotes the octic residue symbol modulo p. The identity (i) gives a generalization of the above equivalence.

We can now state

Theorem 1.5 ([68]). *Let p be any odd prime. Then we have the following arithmetic congruence relation*

$$\#\{x \in \mathbb{F}_p \mid x^4 - 2x^2 + 2 = 0\} = 1 + \left(\frac{-1}{p}\right) + d(p).$$

Corollary. $\mathrm{Spl}\{x^4 - 2x^2 + 2\} = \{p \mid p \equiv 1 \pmod 8, d(p) = 2\}$.

Remark 1.2. The functions $\eta(6\tau)\eta(18\tau)$, $\eta(2\tau)\eta(22\tau)$ and $\eta(8\tau)\eta(16\tau)$ are cusp forms of weight 1 on $\Gamma_0(108)$, $\Gamma_0(44)$ and $\Gamma_0(128)$ respectively. Also Tunnell ([106]) proved that $\eta(8\tau)\eta(16\tau)$ is the unique normalized newform of weight 1, level 128 and character χ_{-2} corresponding to $\mathbb{Q}(\sqrt{-2})$. A. Weil characterized the Dirichlet series corresponding to modular forms for $\Gamma_0(N)$ by functional equations for many associated Dirichlet series ([109]). Its Fourier coefficients are effective to describe the set $\mathrm{Spl}\{f\}$.

Remark 1.3. Let Π be the set of all prime numbers and $T \subset \Pi$ be any subset. For any real $x \geq 1$, we put

$$\delta(x,T) = \frac{\#\{p \in T \,|\, p < x\}}{\#\{p \in \Pi \,|\, p < x\}}.$$

If T is a set of primes such that $\lim_{x\to\infty} \delta(x,T) = \delta(T) < \infty$, then T has density $\delta(T)$. We have now following theorem.

Tchebotarev Density Theorem. *Let $f(x)$ be an irreducible polynomial in $\mathbb{Z}[x]$ with Galois group G, and let C be a fixed conjugacy class of elements in G. Let S be the set of primes p whose Artin symbol C_p equals to C. Then S has a density, and*

$$\delta(S) = \frac{\#C}{\#G}.$$

In particular, if $C = \{1\}$, then $S = \mathrm{Spl}\{f\}$ and $\delta(S) = 1/\#G$.

If $f(x) = x^5 - x - 1$, then the Galois group of $f(x)$ is the symmetric group S_5. Therefore $f(x)$ is one of the non-solvable polynomials. What is the rule to determine the set $\mathrm{Spl}\{x^5 - x - 1\}$? Recall that Wyman ([110]) discussed the relative size of $\mathrm{Spl}\{x^5 - x - 1\}$.

1.2 Modular forms and Hecke operators

In this section we collect without proof some facts from the theory of modular forms that we shall need in this course. We begin with some basic notations.

1.2.1 $\mathrm{SL}_2(\mathbb{Z})$ *and its congruence subgroups*

We put

$$\Gamma = \mathrm{SL}_2(\mathbb{Z}) = \left\{ \begin{pmatrix} a & b \\ c & d \end{pmatrix} : a, b, c, d \in \mathbb{Z},\ ad - bc = 1 \right\}.$$

The group Γ is called the full modular group. Let N be a positive integer. The *principal congruence subgroup of level* N is denoted by $\Gamma(N)$ and consists of all matrices in Γ satisfying

$$\begin{pmatrix} a & b \\ c & d \end{pmatrix} \equiv \begin{pmatrix} 1 & 0 \\ 0 & 1 \end{pmatrix} \pmod N.$$

Since this is the kernel of the natural mapping (reduction mod N)

$$\mathrm{SL}_2(\mathbb{Z}) \to \mathrm{SL}_2(\mathbb{Z}/N\mathbb{Z}),$$

$\Gamma(N)$ is a normal subgroup of finite index in $\mathrm{SL}_2(\mathbb{Z})$. The *Hecke subgroup of level* N is denoted by $\Gamma_0(N)$ and consists of all matrices

$$\begin{pmatrix} a & b \\ c & d \end{pmatrix} \in \Gamma$$

such that $N \mid c$. Since

$$\Gamma(N) \subseteq \Gamma_0(N) \subseteq \Gamma,$$

$\Gamma_0(N)$ has finite index in Γ. The subgroup $\Gamma_1(N)$ consists of all matrices $\gamma \in \Gamma$ satisfying

$$\gamma \equiv \begin{pmatrix} 1 & * \\ 0 & 1 \end{pmatrix} \pmod{N}.$$

Clearly

$$\Gamma(N) \subseteq \Gamma_1(N) \subseteq \Gamma_0(N) \subseteq \Gamma.$$

A *congruence subgroup* of Γ is a subgroup which contains $\Gamma(N)$ for some N. Thus $\Gamma_0(N)$, $\Gamma_1(N)$ are examples of congruence subgroups.

1.2.2 *The upper half-plane*

Let S denote the *upper half-plane*

$$S = \{z = x + iy : x, y \in \mathbb{R}, y > 0\}.$$

Let $\mathrm{GL}_2^+(\mathbb{R})$ be the group of 2 by 2 matrices with real entries and positive determinant. Then $\mathrm{GL}_2^+(\mathbb{R})$ acts on S as a group of holomorphic automorphisms by

$$\gamma : z \mapsto \frac{az + b}{cz + d} \quad \text{for} \quad \gamma = \begin{pmatrix} a & b \\ c & d \end{pmatrix} \in \mathrm{GL}_2^+(\mathbb{R}).$$

Let S^* denote the union of S and the rational numbers $\mathbb{Q}$ together with a symbol ∞. The action of Γ on S can be extended to S^* by defining

$$\begin{pmatrix} a & b \\ c & d \end{pmatrix} \cdot \infty = \frac{a}{c} \quad (c \neq 0),$$

$$\begin{pmatrix} a & b \\ 0 & d \end{pmatrix} \cdot \infty = \infty,$$

and

$$\begin{pmatrix} a & b \\ c & d \end{pmatrix} \cdot \frac{r}{s} = \frac{ar + bs}{cr + ds}$$

for rational number r/s with $\gcd(r, s) = 1$, with the understanding that when $cr + ds = 0$, the right hand side of the above equation is ∞. The rational numbers together with ∞ are called *cusps*.

If G is a discrete subgroup of $\mathrm{SL}_2(\mathbb{R})$, then the orbit space S^*/G can be given the structure of a compact Riemann surface X_G. We will be interested in the case that G is a congruence subgroup of Γ. In that case, the algebraic curve corresponding to X_G is called a *modular curve*. In the case $G = \Gamma(N)$, $\Gamma_1(N)$ or $\Gamma_0(N)$, the corresponding modular curve is denoted by $X(N)$, $X_1(N)$, or $X_0(N)$, respectively.

1.2.3 *Modular forms and cusp forms*

Let f be a holomorphic function on S and k a positive integer. For

$$\gamma = \begin{pmatrix} a & b \\ c & d \end{pmatrix} \in \mathrm{GL}_2^+(\mathbb{R}),$$

define

$$(f|_k\gamma)(z) = (\det(\gamma))^{\frac{k}{2}}(cz + d)^{-k} f\left(\frac{az + b}{cz + d}\right).$$

For fixed k, the mapping $\gamma : f \mapsto f|_k\gamma$ defines an action of $\mathrm{GL}_2^+(\mathbb{R})$ on the space of holomorphic functions on S. Let G be a subgroup of finite index in Γ. Let f be a holomorphic function on S such that $f|_k\gamma = f$ for all $\gamma \in G$. Since G has finite index in Γ,

$$\begin{pmatrix} 1 & 1 \\ 0 & 1 \end{pmatrix}^M = \begin{pmatrix} 1 & M \\ 0 & 1 \end{pmatrix} \in G$$

for some positive integer M. Hence $f(z + M) = f(z)$ for all $z \in S$. So, f has a Fourier expansion at infinity,

$$f(z) = \sum_{n=-\infty}^{\infty} a_n q_M^n \quad \text{with } q_M = e^{\frac{2\pi i z}{M}}.$$

We say that f is *holomorphic at infinity* if $a_n = 0$ for all $n < 0$. We say it *vanishes at infinity* if $a_n = 0$ for all $n \leqq 0$. Let $\sigma \in \Gamma$. Then $\sigma^{-1}G\sigma$ also has finite index in Γ and $(f|_k\sigma)|_k\gamma = f|_k\sigma$ for all $\gamma \in \sigma^{-1}G\sigma$. So for any $\sigma \in \Gamma$, $f|_k\sigma$ also has a Fourier expansion at infinity. We say that f is holomorphic at the cusps if $f|_k\sigma$ is holomorphic at infinity for all $\sigma \in \Gamma$.

Let N be a positive integer and χ a Dirichlet character mod N. A *modular form* on $\Gamma_0(N)$ of type (k, χ) is a holomorphic function f on S such that

(1) $f|_k \begin{pmatrix} a & b \\ c & d \end{pmatrix} = \chi(d) f$ for all $\begin{pmatrix} a & b \\ c & d \end{pmatrix} \in \Gamma_0(N)$, and
(2) f is holomorphic at the cusps.

Note that (1) implies $f|_k\gamma = f$ for all $\gamma \in \Gamma_1(N)$. The Fourier expansion of such a form f is

$$f(z) = \sum_{n=0}^{\infty} a_n q^n, \quad q = e^{2\pi i z}.$$

The integer k is called the weight of f. Such a modular form is called a *cusp form* if it vanishes at the cusps. The modular forms on $\Gamma_0(N)$ of type (k, χ) form a complex linear space $M_k(\Gamma_0(N), \chi)$, and this has as a subspace the set $S_k(\Gamma_0(N), \chi)$ of all cusp forms. The subspace has a canonical complement,

$$M_k(\Gamma_0(N), \chi) = \mathscr{E}_k(\Gamma_0(N), \chi) \oplus S_k(\Gamma_0(N), \chi)$$

and the space $\mathscr{E}_k$ is called the space spanned by *Eisenstein series*. These spaces are finite dimensional.

1.2.4 *Hecke operators*

Let p denotes a prime number and $f(z) = \sum_{n=0}^{\infty} a_n q^n$ be a modular form on $\Gamma_0(N)$ of type (k, χ). The *Hecke operators* T_p and U_p are defined by

$$f|_k T_p(z) = \sum_{n=0}^{\infty} a_{np} q^n + \chi(p) p^{k-1} \sum_{n=0}^{\infty} a_n q^{np} \quad \text{if } p \nmid N,$$

$$f|_k U_p(z) = \sum_{n=0}^{\infty} a_{np} q^n \quad \text{if } p \mid N.$$

It is easy to show that $f|T_p$ and $f|U_p$ are also modular forms on $\Gamma_0(N)$ of type (k, χ), and they are cusp forms if f is a cusp form.

Theorem 1.6 (Hecke-Petersson). *The operators T_p for $p \nmid N$ are commuting linear transformations of $S_k(\Gamma_0(N), \chi)$. The space can be decomposed as a direct sum of common eigenspaces of the operators T_p.*

Let $f \in S_k(\Gamma_0(N), \chi)$. We say that f is an *eigenform* if f is an eigenfunction for all the Hecke operators T_p. If

$$f(z) = \sum_{n=1}^{\infty} a_n e^{2\pi i z}$$

is the Fourier expansion at ∞, and $a_1 = 1$, we call it *normalized.* For the above Fourier expansion, we attach an L-function by

$$L(s,f) = \sum_{n=1}^{\infty} \frac{a_n}{n^s}.$$

Then we have:

Theorem 1.7 (Hecke-Petersson). *The space $S_k(\Gamma_0(N),\chi)$ has a basis of normalized eigenfunctions for all operators T_p. If f is a normalized newform, its Dirichlet series $L(s,f)$ extends to an entire function and has an Euler product expansion*

$$L(s,f) = \prod_{p|N}\left(1 - \frac{a_p}{p^s}\right)^{-1} \prod_{p\nmid N}\left(1 - \frac{a_p}{p^s} + \frac{\chi(p)}{p^{2s+1-k}}\right)^{-1}$$

which converges absolutely for $\mathrm{Re}\, s > (k+2)/2$.

Remark 1.4. Suppose $N' \mid N$ and χ is a Dirichlet character modulo N'. If $f(z) \in S_k(\Gamma_0(N'),\chi)$ and $dN' \mid N$, then $f(dz) \in S_k(\Gamma_0(N),\chi)$. The forms on $\Gamma_0(N)$ which may be obtained in this way from a divisor N' of N ($N' \neq N$), span a subspace of $S_k(\Gamma_0(N),\chi)$ called the space of *oldforms.* Its canonical complement is denoted by $S_k^{\mathrm{new}}(\Gamma_0(N),\chi)$ and the eigenforms in this space are called *newforms.*

Remark 1.5. The space $S_k(\Gamma_1(N),1)$ can be decomposed according to the Dirichlet characters χ mod N which are the characters of $\Gamma_0(N)/\Gamma_1(N)$:

$$S_k(\Gamma_1(N)) = S_k(\Gamma_1(N),1) = \bigoplus_{\chi} S_k(\Gamma_0(N),\chi).$$

The Hecke operators on $S_k(\Gamma_1(N))$ respect the above decomposition of this space. We have a unified definition of the Hecke operators T_m for all positive integers m. For $f(z) = \sum_{n=1}^{\infty} a_n e^{2\pi i n z}$ in $S_k(\Gamma_0(N),\chi)$, the action of T_m is defined by

$$f|_k T_m(z) = \sum_{n=1}^{\infty} b_n q^n \quad (q = e^{2\pi i z}),$$
$$b_n = \sum_{d|\gcd(m,n)} \chi(d) d^{k-1} a_{mn/d^2},$$

where we put $\chi(d) = 0$ whenever $\gcd(d,N) > 1$.

Chapter 2

Hilbert class fields over imaginary quadratic fields

Let K be an imaginary quadratic field, say $K = \mathbb{Q}(\sqrt{-q})$ with a prime number $q \equiv -1 \bmod 8$, and let h be the class number of K. By a classical theory of complex multiplication, the Hilbert class field L of K can be generated by any one of the class invariants over K, which is necessarily an algebraic integer, and a defining equation of which is denoted by $\Phi(x) = 0$. The main purpose of this chapter is to establish the following theorem concerning the arithmetic congruence relation for $\Phi(x)$ ([43]):

Theorem 2.1. *Let p be any prime not dividing the discriminant D_Φ of $\Phi(x)$. Suppose that the ideal class group of K is cyclic. Then we have*

$$\#\{x \in \mathbb{F}_p : \Phi(x) = 0\} = \frac{h}{6}a(p)^2 + \frac{h}{6}a(p) - \frac{1}{2}\left(\frac{-q}{p}\right) + \frac{1}{2},$$

where $\left(\frac{*}{p}\right)$ denotes the Legendre symbol and $a(p)$ denotes the p-th Fourier coefficient of a cusp form which will be defined by (1) in Section 2.2 below. One notes that in case $p = 2$, we have $\left(\frac{-q}{2}\right) = 1$.

2.1 The classical theory of complex multiplication ([21], [31], [113])

Let Λ be a lattice in the complex plane $\mathbb{C}$, and define

$$G_l(\Lambda) = \sum_{\omega \neq 0} \omega^{-l},$$

$$g_2(\Lambda) = 60G_4(\Lambda), \quad g_3(\Lambda) = 140G_6(\Lambda),$$

where l denotes a positive integer and the sum is taken over all non-zero ω in Λ. The torus $\mathbb{C}/\Lambda$ is analytically isomorphic to the elliptic curve E

defined by

$$y^2 = 4x^3 - g_2(\Lambda)x - g_3(\Lambda)$$

via the Weierstrass parametrization

$$\mathbb{C}/\Lambda \ni z \longmapsto (\wp(z), \wp'(z)) \in E,$$

where

$$\wp(z) = \frac{1}{z^2} + \sum_{\omega \neq 0} \left\{ \frac{1}{(z-\omega)^2} - \frac{1}{\omega^2} \right\}, \quad \wp'(z) = \sum_{\omega} \frac{-2}{(z-\omega)^3}.$$

Let Λ and M be two lattices in $\mathbb{C}$. Then the two tori $\mathbb{C}/\Lambda$ and $\mathbb{C}/M$ are isomorphic if and only if there exists a complex number α such that $\Lambda = \alpha M$. If this condition is satisfied, the two lattices Λ and M are said to be linearly equivalent, and we write $\Lambda \sim M$. If so, we have a bijection between the set of lattices in $\mathbb{C}$ modulo $\sim$ and the set of isomorphism classes of elliptic curves. Let us define an *invariant* j depending only on the isomorphism classes of elliptic curves:

$$j(\Lambda) = \frac{1728 g_2^3(\Lambda)}{g_2^3(\Lambda) - 27 g_3^2(\Lambda)}.$$

In fact, $j(\alpha\Lambda) = j(\Lambda)$ for all $\alpha \in \mathbb{C}$. Take a basis $\{\omega_1, \omega_2\}$ of Λ over the ring of rational integers $\mathbb{Z}$ such that $\mathrm{Im}(\omega_1/\omega_2) > 0$ and write $\Lambda = [\omega_1, \omega_2]$. Since $[\omega_1, \omega_2] \sim [\omega_1/\omega_2, 1]$, the invariant $j(\Lambda)$ is determined by $\tau = \omega_1/\omega_2$ which is called the *modulus* of E. Therefore we can write the following: $j(\Lambda) = j(\tau)$. The lattice Λ has many different pairs of generators, the most general pair $\{\omega_1', \omega_2'\}$ with τ' in the upper half-plane having the form

$$\begin{cases} \omega_1' = a\omega_1 + b\omega_2 \\ \omega_2' = c\omega_1 + d\omega_2 \end{cases}$$

with $\begin{pmatrix} a & b \\ c & d \end{pmatrix} \in \mathrm{SL}(2, \mathbb{Z})$, the special linear group of degree 2 with coefficients in $\mathbb{Z}$. Thus the function $j(\tau)$ is a modular function with respect to $\mathrm{SL}(2, \mathbb{Z})$. It is well known that

$$j(\sqrt{-1}) = 1728, \quad j(e^{2\pi\sqrt{-1}/3}) = 0, \quad j(\infty) = \infty.$$

The modular function $j(\tau)$ can be characterized by the above properties.

Let there be given a lattice Λ and the elliptic curve E as described in the above. If for some $\alpha \in \mathbb{C} - \mathbb{Z}$, $\wp(\alpha z)$ is a function on $\mathbb{C}/\Lambda$, then we say that E admits multiplication by α; and then α and ω_1/ω_2 are in the same quadratic field. If E admits multiplication by α_1 and α_2, then E admits multiplication by $\alpha_1 \pm \alpha_2$ and $\alpha_1\alpha_2$. Thus the set of all such α is an order in an imaginary quadratic field K. Consider the case when E admits multiplication by the maximal order $\mathfrak{o}_K$ in K. Then the invariant j defines a function on the ideal classes $k_0, k_1, \ldots, k_{h-1}$ of K (h being the class number of K) and the numbers $j(k_i)$ are called 'singular values' of j. Put

$$A = \left\{ \begin{pmatrix} a & b \\ 0 & d \end{pmatrix} : ad = n > 0,\ 0 \leqq b < d,\ (a,b,d) = 1,\ a,b,d \in \mathbb{Z} \right\},$$

and consider the polynomial

$$F_n(t) = \prod_{\alpha \in A} (t - j(\alpha z)).$$

We may view $F_n(t)$ as a polynomial in two independent variables t and j over $\mathbb{Z}$, and write it as

$$F_n(t) = F_n(t, j) \in \mathbb{Z}[t, j].$$

Let us put $H_n(j) = F_n(j, j)$. Then $H_n(j)$ is a polynomial in j with coefficients in $\mathbb{Z}$, and if n is not a square, then the leading coefficient of $H_n(j)$ is ± 1. This equation

$$H_n(j) = 0$$

is called the *modular equation of order* n. Now we can find an element w in $\mathfrak{o}_K$ such that the norm of w is square-free:

$$w = \begin{cases} 1 + \sqrt{-1}, & \text{if } K = \mathbb{Q}(\sqrt{-1}), \\ \sqrt{-m}, & \text{if } K = \mathbb{Q}(\sqrt{-m}) \text{ with } m > 1 \text{ and square-free.} \end{cases}$$

Let $\{\omega_1, \omega_2\}$ be a basis of an ideal in an ideal class k_i such that $\mathrm{Im}\,(\omega_1/\omega_2) > 0$. Then

$$\begin{cases} w\omega_1 = a\omega_1 + b\omega_2 \\ w\omega_2 = c\omega_1 + d\omega_2 \end{cases}$$

with integers a, b, c, d and the norm of w is equal to $ad - bc$. Thus $\alpha = \begin{pmatrix} a & b \\ c & d \end{pmatrix}$ is primitive and $\alpha\tau = \tau$. Hence $j(\tau) = j(k_i)$ is a root of the modular equation $H_n(j) = 0$. Therefore we have the following

(i) $j(k_i)$ is an algebraic integer.

Furthermore we know

(ii) $K(j(k))$ is the Hilbert class field of K.

By the class field theory, there exists a canonical isomorphism between the ideal class group C_K of K and the Galois group G of $K(j(k_i))/K$, and we have the following formulas which describe how it operates on the generator $j(k_i)$:

(iii) Let σ_k be the element of G corresponding to an ideal class k by the canonical isomorphism. Then

$$\sigma_k(j(k')) = j(k^{-1}k')$$

for any $k' \in C_K$.

(iv) For each prime ideal $\mathfrak{p}$ of K of degree 1, we have

$$j(\mathfrak{p}^{-1}k) \equiv j(k)^{N(\mathfrak{p})} \bmod \mathfrak{p}, \quad k \in C_K,$$

where $N(\mathfrak{p})$ denotes the norm of $\mathfrak{p}$.

(v) The invariants $j(k_i)$, $i = 0, 1, \ldots, h-1$ of K form a complete set of conjugates over the field of rational numbers $\mathbb{Q}$.

2.2 Proof of Theorem 2.1

Let q be a prime number such that $q \equiv -1 \bmod 8$, $K = \mathbb{Q}(\sqrt{-q})$ and let h be the class number of K, which is necessarily odd. For $0 \leqq i \leqq h-1$, we denote by $Q_{k_i}(x, y)$ the binary quadratic form corresponding to the ideal class k_i (k_0: principal class) in K and put

$$\theta_i(\tau) = \frac{1}{2} \sum_{n=0}^{\infty} A_{k_i}(n) e^{2\pi\sqrt{-1}n\tau} \quad (\mathrm{Im}\,(\tau) > 0),$$

where $A_{k_i}(n)$ is the number of integral representations of n by the form Q_{k_i}. Then the following lemma is classical:

Lemma 2.1. *1) If p is any odd prime, except q, then we have*

$$\frac{1}{2} A_{k_0}(p) + \sum_{i=1}^{h-1} A_{k_i}(p) = 1 + \left(\frac{-q}{p}\right).$$

2) If we identify opposite ideal classes by each other, there remain only $A_{k_0}(p)$, $A_{k_1}(p)$, $\ldots$, $A_{k_{(h-1)/2}}(p)$, among which there is at most one non-zero element.

Moreover, for each ideal class k in K, we have

Lemma 2.2. *1)*

$$A_k(n) = 2\#\{\mathfrak{a} \subset \mathfrak{o}_K \,:\, \mathfrak{a} \in k^{-1},\, N(\mathfrak{a}) = n\},$$

2)

$$2A_k(mn) = \sum_{\substack{k_1 k_2 = k \\ k_1, k_2 \in C_K}} A_{k_1}(m) A_{k_2}(n) \quad \textit{if } (m,n) = 1.$$

Let χ be any character ($\neq 1$) on the group C_K of ideal classes and put

$$A(n) = \frac{1}{2} \sum_{k_i \in C_K} \chi(k_i) A_{k_i}(n).$$

Then we have the following multiplicative formulas.

Lemma 2.3. *1)* $A(mn) = A(m)A(n)$ *if* $(m,n) = 1$,
2) $A(p)A(p^r) = A(p^{r+1}) + \left(\frac{-q}{p}\right) A(p^{r-1})$ *for prime* $p(\neq q)$ *and* $r \geqq 1$,

3) $A(qn) = A(q)A(n)$.

We define here two functions f and F as follows:

$$f(\tau) = \theta_0(\tau) - \theta_1(\tau), \tag{2.1}$$

and

$$F(\tau) = \sum_{i=0}^{h-1} \chi(k_i)\theta_i(\tau) = \sum_{n=1}^{\infty} A(n) e^{2\pi\sqrt{-1}n\tau}, \tag{2.2}$$

where $\theta_0(\tau)$ is the *theta-function* corresponding to the principal class k_0. Then $f(\tau)$ is a normalized cusp form on the congruence subgroup $\Gamma_0(q)$ of weight 1 and character $\left(\frac{-q}{p}\right)$, and moreover, by Lemma 2.3, $F(\tau)$ is a normalized newform on $\Gamma_0(q)$ of weight 1 and character $\left(\frac{-q}{p}\right)$ (cf. [35]). From now on, we assume that the ideal class group C_K of K is cyclic. By Lemma 2.1, we shall calculate the Fourier coefficients of $f(\tau)$ and $F(\tau)$. Let

$$C_K = \langle k_1 \rangle \quad \text{and} \quad \chi(k_1) = e^{2\pi\sqrt{-1}/h}.$$

Then we can write the function $F(\tau)$ as

$$F(\tau) = \theta_0(\tau) + 2 \sum_{i=1}^{(h-1)/2} \cos\frac{2\pi i}{h}\theta_i(\tau),$$

where $k_i = k_1^i$ $\left(1 \leqq i \leqq \frac{1}{2}(h-1)\right)$. If $\left(\frac{-q}{p}\right) = -1$, then $A_k(p) = 0$ for all $k \in C_K$. If $\left(\frac{-q}{p}\right) = 1$, then $(p) = \mathfrak{p}\bar{\mathfrak{p}}$ $(\mathfrak{p} \neq \bar{\mathfrak{p}})$ in K, where $\mathfrak{p}$ denotes a prime ideal in K and $\bar{\mathfrak{p}}$ a conjugate of $\mathfrak{p}$. We denote by $k_\mathfrak{p}$ the ideal class such that $\mathfrak{p} \in k_\mathfrak{p}$. If $k_\mathfrak{p}$ is ambiguous, then

$$A_k(p) = \begin{cases} 4, \text{ if } k = k_\mathfrak{p}^{-1}, \\ 0, \text{ otherwise.} \end{cases}$$

If, k is not ambiguous, then

$$A_k(p) = \begin{cases} 2, \text{ if } k = k_\mathfrak{p} \text{ or } k = k_\mathfrak{p}^{-1}, \\ 0, \text{ otherwise.} \end{cases}$$

In the case $p = q$, put $(p) = \mathfrak{p}^2$ $(\mathfrak{p} = \bar{\mathfrak{p}})$ with $\mathfrak{p} \in k_\mathfrak{p}$. Then we have

$$A_k(p) = \begin{cases} 2, \text{ if } k = k_\mathfrak{p}, \\ 0, \text{ otherwise.} \end{cases}$$

Let $a(n)$ be the n-th coefficient of the Fourier expansion for $f(\tau)$:

$$f(\tau) = \sum_{n=1}^{\infty} a(n) e^{2\pi\sqrt{-1}n\tau}.$$

By the above results, we have the following formulas for $a(p)$ and $A(p)$.

Lemma 2.4. *Suppose that the ideal class group C_K of K is cyclic. Then, for each prime p, the Fourier coefficients $a(p)$ and $A(p)$ are given as follows:*

$$a(p) = \begin{cases} 0, & \text{if } \left(\frac{-q}{p}\right) = -1, \\ 2, & \text{if } \left(\frac{-q}{p}\right) = 1 \text{ and } p = x^2 + xy + \frac{1+q}{4}y^2 \ (x, y \in \mathbb{Z}), \\ 0 \text{ or } 1, & \text{if } \left(\frac{-q}{p}\right) = 1 \text{ and } k_\mathfrak{p} \neq k_0 \text{ with } (p) = \mathfrak{p}\bar{\mathfrak{p}},\ \mathfrak{p} \in k_\mathfrak{p}, \\ 1, & \text{if } p = q, \end{cases}$$

and

$$A(p) = \begin{cases} 0, & \text{if } \left(\frac{-q}{p}\right) = -1, \\ 2, & \text{if } \left(\frac{-q}{p}\right) = 1 \text{ and } p = x^2 + xy + \frac{1+q}{4}y^2 \ (x, y \in \mathbb{Z}), \\ 2\cos\frac{2\pi n}{h}, & \text{if } \left(\frac{-q}{p}\right) = 1 \text{ and } k_\mathfrak{p} \neq k_n^{\pm 1} \ (\neq k_0) \text{ with } (p) = \mathfrak{p}\bar{\mathfrak{p}}, \\ & \quad \mathfrak{p} \in k_\mathfrak{p} \ (1 \leqq n \leqq (h-1)/2). \end{cases}$$

Let

$$\Phi(x) = 0$$

be the defining equation of a generating element of the Hilbert class field L over the imaginary quadratic field $K = \mathbb{Q}(\sqrt{-q})$. Then the polynomial $\Phi(x)$ is one of the irreducible factors of the modular polynomial $H_q(x)$. We say simply $\Phi(x)$ is a modular polynomial. Now, in order to prove Theorem 2.1, it is enough to show that if the ideal class group C_K is a cyclic group of order h, then

$$\#\{x \in \mathbb{F}_p \mid \Phi(x) = 0\}$$
$$= \begin{cases} 1, \text{ if } \left(\frac{-q}{p}\right) = -1, \\ h, \text{ if } \left(\frac{-q}{p}\right) = 1 \text{ and } p = x^2 + xy + \frac{1+q}{4}y^2 \ (x, y \in \mathbb{Z}), \\ 0, \text{ if } \left(\frac{-q}{p}\right) = 1 \text{ and } k_{\mathfrak{p}} \neq k_0 \text{ with } (p) = \mathfrak{p}\bar{\mathfrak{p}},\ \mathfrak{p} \in k_{\mathfrak{p}}. \end{cases}$$

We denote by H the ideal group corresponding to the Hilbert class field L of K:

$$H = \{(\alpha) : \textit{principal ideals in } K\}.$$

***Case* 1.** $\left(\frac{-q}{p}\right) = 1$. Let $(p) = \mathfrak{p}\bar{\mathfrak{p}}$ in K. Then we have the following relation:

$$\begin{aligned} \mathfrak{p} \in H &\Longleftrightarrow \mathfrak{p} = (\pi), \quad \pi = a + b\omega \quad (\omega = (1+\sqrt{-q})/2,\ a, b \in \mathbb{Z}) \\ &\Longleftrightarrow p = N(\mathfrak{p}) = a^2 + ab + \frac{1+q}{4}b^2 \quad (a, b \in \mathbb{Z}), \end{aligned}$$

and

$$\mathfrak{p} \text{ splits completely in } L \Longleftrightarrow \Phi(x) \bmod p \text{ has exactly } h \text{ factors.}$$

Therefore

$$p = a^2 + ab + \frac{1+q}{4}b^2 \quad (a, b \in \mathbb{Z}) \Longleftrightarrow \Phi(x) \bmod p \text{ has exactly } h \text{ factors.}$$

On the other hand, it is obvious that

$$\begin{aligned} \mathfrak{p} \notin H &\Longleftrightarrow \mathfrak{p} \text{ is a product of prime ideals of degree} > 1 \text{ in } L \\ &\Longleftrightarrow \Phi(x) \bmod p \text{ has no linear factors in } \mathbb{F}_p[x]. \end{aligned}$$

***Case* 2.** $\left(\frac{-q}{p}\right) = -1$. The polynomial $\Phi(x)$ splits completely modulo p in $\mathfrak{o}_K/(p)$ and the field $\mathfrak{o}_K/(p)$ is a quadratic extension of $\mathbb{F}_p$. Therefore

$$\Phi(x) \bmod p = h_1(x)h_2(x)\cdots h_t(x)$$

and $\deg h_i \leqq 2$ $(i = 1, 2, \ldots, t)$, where each $h_i(x)$ is irreducible in $\mathbb{F}_p[x]$. Since the class number h of K is odd, the number of indices i for which $\deg h_i = 1$ is odd. We shall show that there exists one and only one such i. The *dihedral group* D_h has $2h$ elements and is generated by r, s with the defining relations

$$r^h = s^2 = 1, \quad srs = r^{-1}.$$

Let K_0 be the maximal real subfield of L. We have the following diagram:

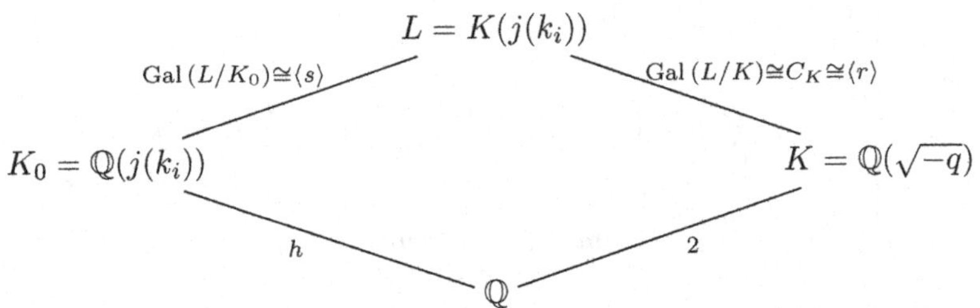

Let $\mathfrak{o}_{K_0}$ be the ring of algebraic integers in K_0. Then the ideal $p\,\mathfrak{o}_{K_0}$ decomposes into a product of distinct prime ideals in K_0:

$$p\,\mathfrak{o}_{K_0} = \mathfrak{p}_1 \cdots \mathfrak{p}_m \mathfrak{g}_1 \cdots \mathfrak{g}_n,$$

where

$$N_{K_0/\mathbb{Q}}(\mathfrak{p}_l) = p\,(1 \leqq l \leqq m) \quad \text{and} \quad N_{K_0/\mathbb{Q}}(\mathfrak{g}_l) = p^2\,(1 \leqq l \leqq n).$$

Moreover, if $\mathfrak{o}_L$ is the ring of algebraic integers in L, then

$$\mathfrak{p}_l \mathfrak{o}_L = \mathfrak{P}_l \quad (1 \leqq l \leqq m),$$

where each $\mathfrak{P}_l$ is a prime ideal in $\mathfrak{o}_L$. On the other hand, the ideal $p\,\mathfrak{o}_L$ has the following decomposition via the field K:

$$p\,\mathfrak{o}_L = \mathfrak{P}_1 \mathfrak{P}_1^r \cdots \mathfrak{P}_1^{r^{h-1}}.$$

Since $\mathfrak{p}_1^s = \mathfrak{p}_1$, we have also $\mathfrak{P}_1^s = \mathfrak{P}_1$. Similarly, $\mathfrak{P}_l^s = \mathfrak{P}_l$ $(2 \leqq l \leqq m)$. However, since h is odd and $srs = r^{-1}$, we deduce

$$\mathfrak{P}_1^{r^i s} = \mathfrak{P}_1^{r^{-i}} \neq \mathfrak{P}_1^{r^i}, \quad (1 \leqq i \leqq h-1).$$

Since $\mathfrak{P}_l = \mathfrak{P}_1^{r^i}$ for some i, we have $m = 1$. This completes the proof of Theorem 2.1. □

Corollary (Higher Reciprocity Law).

$$\mathrm{Spl}\{\Phi(x)\} = \left\{ p \;\middle|\; p \nmid D_\Phi,\ \left(\frac{-q}{p}\right) = 1 \text{ and } a(p) = 2 \right\}.$$

2.3 Schläfli's modular equation

The problem of determining the *modular polynomial* $F_n(t, j)$ explicitly for an arbitrary order n was treated by N. Yui. But, even for $n = 2$, $F_2(t, j)$ has an astronomically long form. We shall use here the Schläfli modular function $h_0(\tau)$ in place of $j(\tau)$:

$$h_0(\tau) = e^{-\pi\sqrt{-1}/24}\frac{\eta((\tau+1)/2)}{\eta(\tau)} = e^{-\pi\sqrt{-1}\tau/24}\prod_{n=1}^{\infty}(1 + e^{(2n-1)\pi\sqrt{-1}\tau}),$$

where η is the *Dedekind eta function.* This function $h_0(\tau)$ is the modular function for the principal congruence subgroup of level 48 and has the following properties:

$$j(\tau) = \frac{\{h_0(\tau)^{24} - 16\}^3}{h_0(\tau)^{24}} \quad \text{and} \quad h_0\left(-\frac{1}{\tau}\right) = h_0(\tau).$$

Lemma 2.5 ([108]). *Let q be any prime number such that $q \equiv -1 \pmod 8$. Then*

1) $\sqrt{2}h_0(\sqrt{-q}) \in \mathbb{Q}(j(\sqrt{-q}))$,

2) $\sqrt{1/2}h_0(\sqrt{-q})$ *is a unit of an algebraic number field.*

Put

$$x = \frac{1}{\sqrt{2}}h_0(\sqrt{-q}).$$

Then, by Lemma 2.5,1), we have

$$\mathbb{Q}(x) = \mathbb{Q}(j(\sqrt{-q})).$$

The defining equation of x is called *the Schläfli modular equation of order* q ([108], §73- §75 and §131).

Example 2.1 ([108]). $n = 47$. Schläfli's modular equation of order 47 is given by

$$x^5 - x^3 - 2x^2 - 2x - 1 = 0.$$

2.4 The case of $q = 47$

Let $\mathfrak{o}_K$ be the principal order of the imaginary quadratic field $K = \mathbb{Q}(\sqrt{-47})$ and put $\mathfrak{o}_K = [1, \omega]$ with $\omega = (1+\sqrt{-47})/2$. The field K has class number 5. Let

$$\begin{aligned} Q_0(x,y) &= x^2 + xy + 12y^2, \\ Q_1(x,y) &= 7x^2 + 3xy + 2y^2, \\ Q_2(x,y) &= 3x^2 - xy + 4y^2, \end{aligned}$$

be the binary quadratic forms corresponding to the ideals $\mathfrak{o}_K$, $[7, 1+\omega]$. $[3, \omega]$, respectively, and let

$$\theta_i(\tau) = \frac{1}{2}\sum_{n=0}^{\infty} A_{Q_i}(n) e^{2\pi\sqrt{-1}n\tau} \quad (i = 0, 1, 2)$$

be the theta-functions belonging to the above binary quadratic forms, respectively, where $A_{Q_i}(n)$ denotes the number of integral representations of n by the form Q_i. By Lemma 2.1, we have easily the following table:

		$A_{Q_0}(p)$	$A_{Q_1}(p)$	$A_{Q_2}(p)$
$\left(\frac{-47}{p}\right) = -1$		0	0	0
$\left(\frac{-47}{p}\right) = 1$	$p = x^2 + 47y^2$	4	0	0
	$7p = x^2 + 47y^2$	0	2	0
	$3p = x^2 + 47y^2$	0	0	2

For $p = 2, 47$, we have

$$\begin{aligned} &A_{Q_0}(2) = A_{Q_2}(2) = 0, \quad A_{Q_1}(2) = 2; \\ &A_{Q_0}(47) = 2, \quad A_{Q_1}(47) = A_{Q_2}(47) = 0. \end{aligned}$$

Now we define two functions as follows:

$$F_1(\tau) = \theta_0(\tau) - \theta_1(\tau) = \sum_{n=1}^{\infty} a(n) e^{2\pi\sqrt{-1}n\tau},$$

$$F_2(\tau) = \theta_0(\tau) - \theta_2(\tau).$$

Then $F_1(\tau)$ and $F_2(\tau)$ are normalized cusp forms on the group $\Gamma_0(47)$ of weight 1 and character $\left(\frac{-47}{p}\right)$. Put $\varepsilon_0 = \frac{1}{2}(1+\sqrt{5})$ and define

$$\begin{aligned} F_3(\tau) &= \bar{\varepsilon}_0 F_1(\tau) + \varepsilon_0 F_2(\tau) = F_1(\tau) + \varepsilon_0 \eta(\tau)\eta(47\tau) \\ &= \sum_{n=1}^{\infty} A(n) e^{2\pi\sqrt{-1}n\tau}. \end{aligned}$$

Then the function $F_3(\tau)$ is also a normalized cusp form of weight 1 and character $\left(\frac{-47}{p}\right)$ on the group $\Gamma_0(47)$, and the Fourier coefficient $A(n)$ is multiplicative. The Fourier coefficients of $F_1(\tau)$ and $F_2(\tau)$ are obtained by the above table as follow, respectively. For each prime p ($\neq 2, 47$), we have

$$a(p) = \begin{cases} 0 \text{ if } \left(\frac{-47}{p}\right) = -1, \\ 2 \text{ if } \left(\frac{-47}{p}\right) = 1 \text{ and } p = x^2 + 47y^2 \ (x, y \in \mathbb{Z}), \\ 0 \text{ if } \left(\frac{-47}{p}\right) = 1 \text{ and } 3p = x^2 + 47y^2 \ (x, y \in \mathbb{Z}), \\ -1 \text{ if } \left(\frac{-47}{p}\right) = 1 \text{ and } 7p = x^2 + 47y^2 \ (x, y \in \mathbb{Z}), \end{cases} \tag{2.3}$$

and

$$A(p) = \begin{cases} 0 \text{ if } \left(\frac{-47}{p}\right) = -1, \\ 2 \text{ if } \left(\frac{-47}{p}\right) = 1 \text{ and } p = x^2 + 47y^2 \ (x, y \in \mathbb{Z}), \\ -\varepsilon_0 \text{ if } \left(\frac{-47}{p}\right) = 1 \text{ and } 3p = x^2 + 47y^2 \ (x, y \in \mathbb{Z}), \\ -\bar{\varepsilon}_0 \text{ if } \left(\frac{-47}{p}\right) = 1 \text{ and } 7p = x^2 + 47y^2 \ (x, y \in \mathbb{Z}). \end{cases} \tag{2.4}$$

Furthermore we have $a(2) = -1$, $a(47) = A(47) = 1$ and $A(2) = -\bar{\varepsilon}_0$.

Put $h_0(-47) = \sqrt{2}x$. Then the class invariant x satisfies the following Schläfli's modular equation of order 47 (cf. Section 2.3):

$$f_W(x) = x^5 - x^3 - 2x^2 - 2x - 1 = 0 \quad (D_{f_W} = 47^2). \tag{2.5}$$

Let L be the Hilbert class field over K. Then the field L is a splitting field for the polynomial

$$f_H(x) = x^5 - 2x^4 + 2x^3 - 3x^2 - 3x + 6x - 5 = 0 \quad (D_{f_H} = 11^2 \cdot 47^2), \tag{2.6}$$

and the Galois group $G(L/\mathbb{Q})$ is equal to the dihedral group D_5 ([32], [33]).

Put

$$\eta_0 = \frac{1}{2}\left(\frac{47-5\sqrt{5}}{2} + \frac{-5+\sqrt{5}}{2}\sqrt{47\sqrt{5}\varepsilon_0}\right)$$

and

$$\omega_0 = \frac{9353+422\sqrt{5}}{7} - \frac{715+325\sqrt{5}}{4}\sqrt{47\sqrt{5}\varepsilon_0},$$

then from Hasse's result ([32]) we deduce that

$$\theta_H = \frac{1}{5}\left(\sqrt[5]{\omega_0} - \frac{1}{\sqrt[5]{\omega_0}} - \frac{\sqrt[5]{\omega_0^2}}{\eta_0} + \frac{\eta_0}{\sqrt[5]{\omega_0^2}} + 2\right)$$

generates L/K. Consider the following equation ([24], p.492):

$$f_F(x) = x^5 - x^4 + x^3 + x^2 - 2x + 1 = 0. \tag{2.7}$$

It is known that there are two relations

$$\begin{cases} \theta_H = 5\theta_W^2 - 5\theta_W - 2, \\ \theta_W = -\theta_F^4 - 2\theta_F + 1 \end{cases} \tag{2.8}$$

for the real roots θ_W, θ_H and θ_F of (2.5), (2.6) and (2.7), respectively ([111]). Put

$$f_M(x) = x^5 - 2x^4 + 3x^3 + x^2 - x - 1.$$

The discriminant of our polynomial $f_M(x)$ is $5^2 \cdot 47^2$. By a simple calculation, we can verify the following remarkable relation:

$$x^2 - ax + b \mid f_F(x) \Longleftrightarrow f_H(a)f_M(a) = 0, \tag{2.9}$$

where a and b denote any constants. If θ is the real root of the equation $f_M(x) = 0$, then we obtain the following relations by making use of Newton's method:

$$\begin{cases} \theta_H = 2\theta_F^4 + \theta_F^3 + 2\theta_F - 2, \text{(by (2.8))} \\ \theta = -2\theta_F^4 + \theta_F^3 - \theta_F^2 - 3\theta_F + 3, \\ \theta_F = \dfrac{-1}{11}(\theta_H^4 + \theta_H^3 + 5\theta_H^2 + \theta_H - 2), \\ \theta = \dfrac{-1}{11}(\theta_H^4 + \theta_H^3 + 5\theta_H^2 - \theta_H + 9), \\ \theta_F = \dfrac{1}{5}(\theta^4 - 5\theta^3 + 8\theta^2 - 8\theta - 2), \\ \theta_H = \dfrac{1}{5}(-\theta^4 + 5\theta^3 - 8\theta^2 + 3\theta + 7). \end{cases} \tag{2.10}$$

Now we consider $f_F(x) \bmod p$ for any odd prime number $p (\neq 47)$. Because of (2.9) and (2.10), the reduced polynomial $f_F \bmod p$ $(p \neq 5, 11)$ can factor over the p-element field $\mathbb{F}_p$ in one of three ways:

1) Five linear factors,
2) (linear)(Quadratic)(Quadratic),
3) Quintic.

The reduced polynomials $f_F \bmod 5$ and $f_F \bmod 11$ have the above type 2). When we combine these with (2.3), we are led to the another proof of the arithmetic congruence relation in the case of $q = 47$ related to Theorem 2.1.

Theorem 2.2. *Let p be any prime, except 47. Let $a(n)$ be the n-th coefficient of the expansion*

$$F_1(\tau) = \sum_{n=1}^{\infty} a(n) e^{2\pi\sqrt{-1}n\tau}.$$

Then the following congruence relation for $f_F(x)$ holds:

$$\#\{x \in \mathbb{F}_p | f_F(x) = 0\} = \frac{5}{6}a(p)^2 + \frac{5}{6}a(p) - \frac{1}{2}\left(\frac{-47}{p}\right) + \frac{1}{2},$$

where for $p = 2$, we understand $\left(\frac{-47}{2}\right) = 1$.

Proof. In order to prove this, it is enough to show the following fact. Let L_p be a splitting field of $f_F(x) \bmod p$ over the field $\mathbb{F}_p$. Then it can be seen that

$$\left(\frac{-47}{p}\right) = -1 \Longleftrightarrow [L_p : \mathbb{F}_p] = 2$$

$$\Longleftrightarrow f_F \bmod p \text{ has exactly one linear factor over } \mathbb{F}_p$$

$$\Longleftrightarrow f_F \bmod p \text{ can factor in type 2).}$$

□

Remark 2.1. Let p be a prime, except 5, 11, 47. Then, by the relation (2.10), $f_F \bmod p$, $f_H \bmod p$, $f_W \bmod p$ and $f_M \bmod p$ can factor over $\mathbb{F}_p$ in the same way. Using Fourier coefficient of $F_2(\tau)$, we have also the same arithmetic congruence relation for $f_F(x)$. On the other hand, using Fourier coefficients $A(p)$ of $F_3(\tau)$ (cf. (2.4)), we have the following relation:

$$\#\{x \in \mathbb{F}_p | f_F(x) = 0\} = A(p)^2 + A(p) - \left(\frac{-47}{p}\right).$$

Finally the following higher reciprocity law for the *Fricke polynomial* $f_F(x)$ holds:

Corollary 2.1.

$$\mathrm{Spl}\{f_F(x)\} = \left\{ p \;\middle|\; \left(\frac{-47}{p}\right) = 1 \text{ and } a(p) = 2 \right\}.$$

Example 2.2. The dihedral group D_h has $(h+3)/2$ conjugate classes:

$$\{1\}, \quad \{sr^i | 1 \leqq i \leqq h\}, \quad \{r^j, r^{-j}\}, \quad j = 1, 2, \ldots, (h-1)/2.$$

Thus we have $(h-1)/2$ irreducible representations of degree 2. Among them, here we consider the representation ρ given by the following

$$\rho(r) = \begin{pmatrix} \varepsilon & 0 \\ 0 & \varepsilon^{-1} \end{pmatrix}, \quad \rho(s) = \begin{pmatrix} 0 & 1 \\ 1 & 0 \end{pmatrix},$$

where $e^{2\pi\sqrt{-1}/h}$. The corresponding character is given by the following

	$\{1\}$	$\{r^j, r^{-j}\}$	$\{sr^i \mid 1 \leqq i \leqq h\}$
ρ	2	$2\cos\frac{2\pi j}{h}$	0

$$j = 1, 2, \ldots, \frac{h-1}{2}.$$

Let $\phi(s)$ be the Dirichlet series associated to the newform $F(\tau)$ (cf. (2.2) in Section 2.1) via the Mellin transform. Since the function $F(\tau)$ is an eigenfunction of all the Hecke operators T_p, U_p, the Dirichlet series $\phi(s)$ has the following Euler product:

$$\begin{aligned}
\phi(s) &= \sum_{n=1}^{\infty} A(n)n^{-s} = (1 - A(q)q^{-s})^{-1} \prod_{p \neq q} \left(1 - A(p)p^{-s} + \left(\frac{-q}{p}\right) p^{-2s}\right)^{-1} \\
&= (1 - q^{-s})^{-1} \prod_{\left(\frac{-q}{p}\right) = -1} (1 - p^{-2s})^{-1} \prod_{p \in P_1} (1 - 2p^{-s} + p^{-2s})^{-1} \\
&\quad \times \prod_{p \in P_2} \left(1 + 2\cos\frac{2\pi n}{h} p^{-s} + p^{-2s}\right)^{-1},
\end{aligned}$$

where

$$P_1 = \left\{ p \;\middle|\; \left(\frac{-q}{p}\right) = 1, \ p = x^2 + xy + \frac{1+q}{4} y^2 \right\},$$

and

$$P_2 = \left\{ p \;\middle|\; \left(\frac{-q}{p}\right) = 1, \ p = \mathfrak{p}\bar{\mathfrak{p}}, \ \mathfrak{p} \neq \text{principal}, \ \mathfrak{p} \in k_n \right\} \cup \{2\}.$$

Let L be the Hilbert class field of the imaginary quadratic field K, and assume that the Galois group $G(L/K)$ is a cyclic group of order h. Then

$L/\mathbb{Q}$ is a non-abelian Galois extension with D_h as Galois group. Let p be any prime number and σ_p a Frobenius map of p in L, and put

$$A_p = \frac{1}{e} \sum_{\alpha \in T} \rho(\sigma_p \alpha),$$

where T is the inertia group of p and $\#T = e$. Then, for the Galois extension $L/\mathbb{Q}$, the Artin L-function is defined by

$$L(s, \rho, L/\mathbb{Q}) = \prod_p \det\left(\begin{pmatrix} 1 & 0 \\ 0 & 1 \end{pmatrix} - A_p N(p)^{-s}\right)^{-1}, \quad \mathrm{Re}\,(s) > 1.$$

A prime p factorizes in L in one of the following ways:

***Case* 1.** $\left(\frac{-q}{p}\right) = 1$. Decomposition field $= K_0$, $\sigma_p = s$, $A_p = \begin{pmatrix} 0 & 1 \\ 1 & 0 \end{pmatrix}$.

***Case* 2.** $p \in P_1$. Decomposition field $= L$, $\sigma_p = 1$, $A_p = \begin{pmatrix} 1 & 0 \\ 0 & 1 \end{pmatrix}$.

***Case* 3.** $p \in P_2$. Decomposition field $= K$, If $(p) = \mathfrak{p}\bar{\mathfrak{p}}$ with $\mathfrak{p} \in k_n^{-1}$, then $\sigma_p = r^n$ and $A_p = \begin{pmatrix} \varepsilon^n & 0 \\ 0 & \varepsilon^{-n} \end{pmatrix}$.

***Case* 4.** $p = q$. Ramification exponent $= 2$.

$$\sigma_q = 1, \quad A_q = \frac{1}{2}(\rho(1) + \rho(s)) = \frac{1}{2}\begin{pmatrix} 1 & 1 \\ 1 & 1 \end{pmatrix}.$$

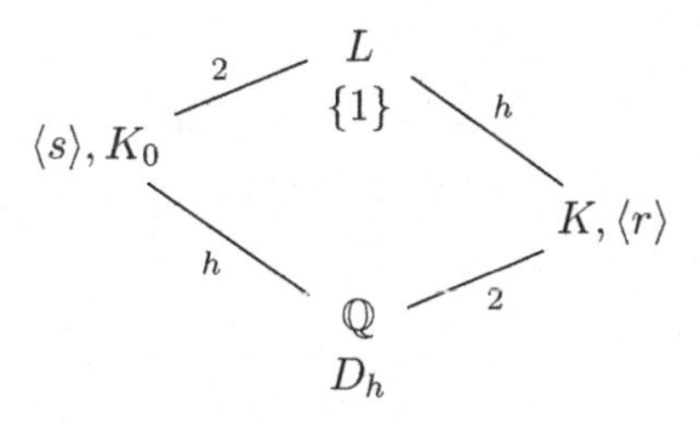

In order to have the explicit form of $L(s, \rho, L/\mathbb{Q})$, we use the above results and obtain

$$\begin{aligned} L(s, \rho, L/\mathbb{Q}) &= \prod_p \det\left(\begin{pmatrix} 1 & 0 \\ 0 & 1 \end{pmatrix} - A_p N(p)^{-s}\right)^{-1} \\ &= \det\left(\begin{pmatrix} 1 & 0 \\ 0 & 1 \end{pmatrix} - q^{-s}\frac{1}{2}\begin{pmatrix} 1 & 1 \\ 1 & 1 \end{pmatrix}\right)^{-1} \prod_{\left(\frac{-q}{p}\right)=-1} \det\left(\begin{pmatrix} 1 & 0 \\ 0 & 1 \end{pmatrix} - p^{-s}\begin{pmatrix} 0 & 1 \\ 1 & 0 \end{pmatrix}\right)^{-1} \\ &\times \prod_{p \in P_1} \det\left(\begin{pmatrix} 1 & 0 \\ 0 & 1 \end{pmatrix} - p^{-s}\begin{pmatrix} 1 & 0 \\ 0 & 1 \end{pmatrix}\right)^{-1} \prod_{p \in P_2} \det\left(\begin{pmatrix} 1 & 0 \\ 0 & 1 \end{pmatrix} - p^{-s}\begin{pmatrix} \varepsilon^n & 0 \\ 0 & \varepsilon^{-n} \end{pmatrix}\right)^{-1}. \end{aligned}$$

It is clear that above Euler product, compared with the Euler product of $\phi(s)$, proves the following:

$$L(s, \rho, L/\mathbb{Q}) = \phi(s).$$

Chapter 3

Indefinite modular forms

As shown in Chapters 1 and 2, there are deep relations between the class fields over the imaginary quadratic fields and cusp forms of weight 1. In the first half of this chapter, we study a similar problem for class fields over real quadratic field which satisfies a condition due to Shintani ([97]). In Section 3.1 we recall the definition of Hecke's indefinite modular forms of weight 1 which are associated to real quadratic fields ([34],[35],[65]). In Section 3.2 we summarize certain results of Shintani for the real quadratic problem which is transferable to the imaginary quadratic situation ([97]). In Section 3.3 we apply the result of Shintani to our problem and obtain the three representations for some dihedral cusp forms of weight 1 by positive definite theta series and indefinite theta series. Kac and Peterson in [58] gave many examples of new identities for cusp forms of weight 1 which arise from the Dedekind eta function. In Section 3.4 we shall reconstruct these examples from our point of view, by using the results of Section 3.3. In Section 3.5 we establish the higher reciprocity law for a defining equation of ray class fields over some real quadratic fields.

The second half of this chapter will be devoted to study a relation between quartic residuacity and Fourier coefficients of cusp forms of weight 1 ([42]). Let m be a positive square free integer and ε_m denote the fundamental unit of $\mathbb{Q}(\sqrt{m})$. We consider only those m for which ε_m has norm $+1$. If l is an odd prime such that $\left(\frac{m}{l}\right) = \left(\frac{\varepsilon_m}{l}\right) = 1$, we can ask for the value of the quartic residue symbol $\left(\frac{\varepsilon_m}{l}\right)_4$. Let K be the Galois extension of degree 16 over $\mathbb{Q}$ generated by $\sqrt{-1}$ and $\sqrt[4]{\varepsilon_m}$. Then its Galois group $G(K/\mathbb{Q})$ has just two irreducible representations of degree 2. We can define a cusp form of weight 1 by these representations, which will be denoted by $\Theta(\tau;\, K)$ and we shall show that $\Theta(\tau;\, K)$ has three expressions by definite and indefinite theta series and that the value of the symbol $\left(\frac{\varepsilon_m}{l}\right)_4$ is expressed by the l-th Fourier coefficient of $\Theta(\tau;\, K)$. These results offer us new criteria for ε_m to be a quartic residue modulo l.

3.1 Hecke's indefinite modular forms of weight 1

Let F be a real quadratic field with discriminant D, and $\mathfrak{o}_F$ the ring of all integers in F. Let Q be a natural number and denote by $\mathfrak{U}_0$ the group of totally positive unit ε of $\mathfrak{o}_F$ such that $\varepsilon \equiv 1 \bmod Q\sqrt{D}$. Let $\mathfrak{a}$ be an integral ideal of $\mathfrak{o}_F$, and put $|N(\mathfrak{a})| = A$. Then the *Hecke modular form* for the ideal $\mathfrak{a}$ is defined by

$$\vartheta_\kappa(\tau;\, \rho,\, \mathfrak{a},\, Q\sqrt{D}) = \sum_{\substack{\mu\in\mathfrak{o}_F \\ \mu\equiv\rho \bmod \mathfrak{a}Q\sqrt{D} \\ \mu\in\mathfrak{o}_F/\mathfrak{U}_0,\, N(\mu)\kappa>0}} (\operatorname{sgn}\mu) q^{N(\mu)/AQD},$$

where $\kappa = \pm 1$, $\rho \in \mathfrak{a}$, $\operatorname{Im}(\tau) > 0$ and $q = e^{2\pi i\tau}$. This is a holomorphic function of τ and satisfies

$$\begin{aligned}\vartheta_\pm&\left(\frac{a\tau+b}{c\tau+d};\, \rho,\, \mathfrak{a},\, Q\sqrt{D}\right)\\ &= \left(\frac{D}{|d|}\right) e^{\mp 2\pi i ab\rho\rho'/AQD}(c\tau+d)\vartheta_\pm(\tau;\, a\rho,\, \mathfrak{a},\, Q\sqrt{D})\end{aligned}$$

for all $\begin{pmatrix} a & b \\ c & d \end{pmatrix} \in \Gamma_0(QD)$ ([34],[35]).[1] Therefore $\vartheta_\pm$ is the cusp form of weight 1 for a certain congruence subgroup of level QD under the condition $\vartheta_\pm \not\equiv 0$. If in particular $\mathfrak{a} = \mathfrak{o}_F$, we put

$$\vartheta_\pm(\tau;\, \rho,\, Q\sqrt{D}) = \vartheta_\pm(\tau;\, \rho,\, \mathfrak{o}_F,\, Q\sqrt{D}).$$

3.2 Ray class fields over real quadratic fields

Let there be given a real quadratic field F as described in Section 3.1. Let $\mathfrak{f}$ be a self conjugate integral ideal of $\mathfrak{o}_F$ which satisfies the condition:

$$\text{For any totally positive unit } \varepsilon \text{ of } \mathfrak{o}_F,\ \varepsilon + 1 \notin \mathfrak{f}. \tag{3.1}$$

We denote by $H_F(\mathfrak{f})$ the narrow ray class group modulo $\mathfrak{f}$ of F. Then, under the condition (3.1), the group $H_F(\mathfrak{f})$ has a character χ of the following type:

$$\chi((x)) = \operatorname{sgn} x \quad \text{or} \quad \chi((x)) = \operatorname{sgn} x'$$

[1] For a general treatment of this function via Weil representation, see [58] and [65].

for $x - 1 \in \mathfrak{f}$, where x' denotes the conjugate of x. We denote the *Hecke L-function* of F attached to χ by

$$L_F(s,\chi) = \sum_{c \in H_F(\mathfrak{f})} \chi(c) \sum_{\substack{\mathfrak{a} \in c \\ \mathfrak{a} \subset \mathfrak{o}_F}} N(\mathfrak{a})^{-s} \quad (\mathrm{Re}\,(s) > 1).$$

Then the *Γ-factor* in the functional equation of $L_F(s,\chi)$ is of the form

$$\Gamma\left(\frac{s}{2}\right)\Gamma\left(\frac{s+1}{2}\right).$$

We put

$$H_F(\mathfrak{f})_0 = \{c \in H_F(\mathfrak{f}) \,|\, c' = c\},$$

and assume that

$$[H_F(\mathfrak{f}) : H_F(\mathfrak{f})_0] = 2. \tag{3.2}$$

Let $K_F(\mathfrak{f})$ denote the maximal narrow ray class field over F corresponding to $H_F(\mathfrak{f})$ and σ denote the Artin canonical isomorphism given by class field theory. Let L be the subfield of $\sigma(H_F(\mathfrak{f})_0)$-fixed elements of $K_F(\mathfrak{f})$. Then, under the assumption (3.2), L is a composition of F with a suitable imaginary quadratic field k, and $K_F(\mathfrak{f})$ is an abelian extension of k ([97]).

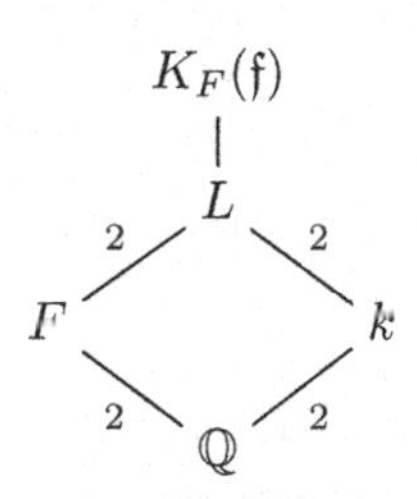

Therefore there exists an integral ideal $\mathfrak{c}$ of k such that $K_F(\mathfrak{f})$ is a class field over k with conductor $\mathfrak{c}$. Let $\mathfrak{f}_\chi$ be the conductor of χ and $\tilde{\chi}$ the primitive character of $H_F(\mathfrak{f}_\chi)$ corresponding to χ. We denote by ξ_χ one of the characters of the group $H_k(\mathfrak{c})$ determined by χ in a natural manner. Let $\mathfrak{c}_\chi$ be the conductor of ξ_χ and $\tilde{\xi}_\chi$ the primitive character of $H_k(\mathfrak{c}_\chi)$ corresponding to ξ_χ. Then we have the following coincidence of two L-functions associated with the real quadratic field F and the imaginary quadratic field k ([97]):

$$L_F(s,\tilde{\chi}) = L_k(s,\tilde{\xi}_\chi). \tag{3.3}$$

3.3 Positive definite and indefinite modular forms of weight 1

In this section we use the same symbols as in Section 3.2. We put

$$K = K_F(\mathfrak{f});$$

and assume that K/k is a cyclic extension. We denote by $\vartheta(F/\mathbb{Q})$ and $\vartheta(k/\mathbb{Q})$ the different of F over $\mathbb{Q}$ and that of k over $\mathbb{Q}$, respectively. Then we have the following relation between the conductor $\mathfrak{c}$ of the cyclic extension K/k and the finite part $\mathfrak{f}$ for the conductor of the abelian extension K/F by Hasse's theorem:

Lemma 3.1. $\mathfrak{f} \cdot \vartheta(F/\mathbb{Q}) = \mathfrak{c} \cdot \vartheta(k/\mathbb{Q})$ *as ideals in* L.

Let us, temporarily, assume that $K/\mathbb{Q}$ is a dihedral extension. Then the Galois group $G(K/\mathbb{Q})$ is the dihedral group D_4 of order 8 and we have the following diagram of fields:

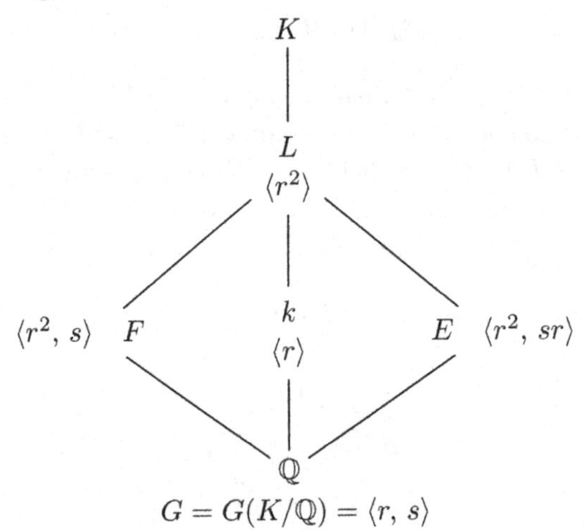

$$G = G(K/\mathbb{Q}) = \langle r,\, s\rangle$$

Here E denotes the imaginary quadratic field determined by F and k. The conductor $\mathfrak{c}$ of K/k is an ideal of $\mathbb{Z}$ by Satz 7 of Halter-Koch ([61]). Now we put

$$\mathfrak{c} = (c), \quad c \in \mathbb{Z}.$$

Since, $\mathfrak{f}' = \mathfrak{f}$, $(\mathfrak{f} \cdot \vartheta(F/\mathbb{Q}))^2$ is an ideal of $\mathbb{Z}$, i.e.,

$$(\mathfrak{f} \cdot \vartheta(F/\mathbb{Q}))^2 = (q^2 \cdot d),$$

where q is a positive integer and d is a positive square-free integer. K/k being a cyclic extension by assumption, we have the following by Lemma 3.1.

Lemma 3.2. $c = q \cdot e_d^{-1}$ *and* $k = \mathbb{Q}(\sqrt{-d})$, *where*

$$e_d = \begin{cases} 1 \text{ if } d \equiv 3 \pmod 4, \\ 2 \text{ otherwise.} \end{cases}$$

We are going to discuss how to obtain an identity between cusp forms of weight 1. Take an integer μ of F such that $\mu < 0$, $\mu' > 0$ and $\mu \equiv 1 \operatorname{mod} \mathfrak{f}$, and denote by the same letter μ the ray class modulo $\mathfrak{f}$ represented by the principal ideal (μ). Then, by the condition (3.1), μ is an element of order 2 of $H_F(\mathfrak{f})$, and by the condition (3.2), we have

$$H_F(\mathfrak{f}) = H_F(\mathfrak{f})_0 + H_F(\mathfrak{f})_0\mu.$$

Let $\langle \mu\mu' \rangle$ be the subgroup of $H_F(\mathfrak{f})_0$ generated by $\mu\mu'$ and let R be a complete set of representatives of $H_F(\mathfrak{f})_0 \operatorname{mod} \langle \mu\mu' \rangle$. Since $\langle \mu\mu' \rangle$ is the subgroup of order 2 of $H_F(\mathfrak{f})_0$, we have

$$H_F(\mathfrak{f}) = R \cup R\mu \cup R\mu' \cup R\mu\mu' \quad \text{(disjoint)}.$$

For $c \in H_F(\mathfrak{f})$, we put

$$\zeta_F(s, c) = \sum_{\substack{\mathfrak{a} \in c \\ \mathfrak{a} \subset \mathfrak{o}_F}} N(\mathfrak{a})^{-s}.$$

Then it is easily checked that

$$\zeta_F(s, \sigma\mu) = \zeta_F(s, \sigma\mu')$$

for $\sigma \in R$. Let χ be a character of $H_F(\mathfrak{f})$ with conductor $\mathfrak{f}(\infty_1)$ satisfying the condition (3.1). Then the Hecke L-function of F attached to χ has the following expression

$$\begin{aligned} L_F(s, \chi) &= \sum_{\sigma \in R} \chi(\sigma)\{\zeta_F(s, \sigma) - \zeta_F(s, \sigma\mu) + \zeta_F(s, \sigma\mu') - \zeta_F(s, \sigma\mu\mu')\} \\ &= \sum_{\sigma \in R} \chi(\sigma)\{\zeta_F(s, \sigma) - \zeta_F(s, \sigma\mu\mu')\}. \end{aligned}$$

Let σ be an element of R and let $\mathfrak{a}_\sigma$ be an integral ideal of σ^{-1}. We put

$$\begin{aligned} A_\sigma^+ &= \{\alpha \in \mathfrak{a}_\sigma \mid \alpha \equiv 1 \operatorname{mod} \mathfrak{f},\ \alpha > 0,\ \alpha' > 0\}, \\ A_\sigma^- &= \{\alpha \in \mathfrak{a}_\sigma \mid \alpha \equiv 1 \operatorname{mod} \mathfrak{f},\ \alpha < 0,\ \alpha' < 0\} \end{aligned}$$

and

$$A_\sigma = A_\sigma^+ \cup A_\sigma^-.$$

Then it is easy to verify that

$$A_\sigma = \{\alpha \in \mathfrak{o}_F \,|\, \alpha \equiv \rho_\sigma \bmod \mathfrak{a}_\sigma \mathfrak{f},\, N(\alpha) > 0\},$$

where ρ_σ denotes an element of $\mathfrak{a}_\sigma$ such that $\rho_\sigma \equiv 1 \bmod \mathfrak{f}$. Moreover, we have the following two bijections:

$$A_\sigma^+ \bmod E_\mathfrak{f}^+ \ni \alpha \bmod E_\mathfrak{f}^+ \Longleftrightarrow \alpha\mathfrak{a}_\sigma^{-1} \in \sigma \cap \mathfrak{o}_F$$

and

$$A_\sigma^- \bmod E_\mathfrak{f}^+ \ni \alpha \bmod E_\mathfrak{f}^+ \Longleftrightarrow \alpha\mathfrak{a}_\sigma^{-1} \in \sigma\mu\mu' \cap \mathfrak{o}_F,$$

where

$$E_\mathfrak{f}^+ = \{\varepsilon :\ \text{unit of } \mathfrak{o}_F \,|\, \varepsilon \equiv 1 \bmod \mathfrak{f},\, \varepsilon > 0,\, \varepsilon' > 0\}.$$

From these correspondences, it is easy to see that

$$\zeta_F(s,\sigma) = \sum_{\alpha \in A_\sigma^+ \bmod E_\mathfrak{f}^+} (N(\alpha)/N(\mathfrak{a}_\sigma))^{-s}$$

and

$$\zeta_F(s,\sigma\mu\mu') = \sum_{\alpha \in A_\sigma^- \bmod E_\mathfrak{f}^+} (N(\alpha)/N(\mathfrak{a}_\sigma))^{-s}.$$

Hence we obtain explicit form of $L_F(s,\chi)$:

$$\begin{aligned} L_F(s,\chi) &= \sum_{\sigma \in R} \chi(\sigma) \sum_{\alpha \in A_\sigma \bmod E_f^+} (\operatorname{sgn}\alpha)(N(\alpha)/N(\mathfrak{a}_\sigma))^{-s} \\ &= \sum_{\sigma \in R} \chi(\sigma) \sum_{\alpha} (\operatorname{sgn}\alpha)(N(\alpha)/N(\mathfrak{a}_\sigma))^{-s}, \end{aligned}$$

where α in the summation runs over all integers of F such that $\alpha \equiv \rho_\sigma \bmod \mathfrak{a}_\sigma\mathfrak{f}$, $\alpha \bmod E_\mathfrak{f}^+$ and $N(\alpha) > 0$. We apply the inverse Mellin transformation on the above L-function and obtain the following indefinite cusp form of weight 1:

$$\begin{aligned} \theta_F(\tau) &= \sum_{\sigma \in R} \chi(\sigma) \sum_{\alpha} (\operatorname{sgn}\alpha) q^{N(\alpha)/N(\mathfrak{a}_\sigma)} \qquad (q = e^{2\pi i \tau}) \\ &= \sum_{\sigma \in R} \chi(\sigma)\theta(QD_1\tau;\, \rho_\sigma,\, \mathfrak{a}_\sigma,\, \mathfrak{f}), \end{aligned}$$

where $\mathfrak{f} = Q\mathfrak{f}_1$, $\mathfrak{f}_1 \mid \sqrt{D}$, $D_1 = N(\mathfrak{f}_1)$ and

$$\theta(\tau;\, \rho_\sigma,\, \mathfrak{a}_\sigma,\, \mathfrak{f}) = \sum_{\alpha} (\operatorname{sgn}\alpha) q^{N(\alpha)/N(\mathfrak{a}_\sigma)QD_1}.$$

In particular, if we put $\mathfrak{f}_1 = \sqrt{D}$, then the above function θ is just the Hecke indefinite modular form defined in Section 3.1.

On the other hand, since K/k is a cyclic extension, we can put

$$H_k(\mathfrak{c})/C = \langle\lambda\rangle,$$

where C denotes the subgroup of $H_k(\mathfrak{c})$ corresponding to K. The generator λ is an element of order $4m$. The restriction of the representation of Gal $(K/\mathbb{Q})$ induced from χ to Gal (K/k) is a direct sum of two distinct primitive characters ξ and ξ' of $H_k(\mathfrak{c})/C$ via the Artin map. Then we consider Hecke L-function attached to ξ:

$$\begin{aligned} L_k(s,\xi) &= \sum_{\mathfrak{a}\subset\mathfrak{o}_k} \xi(\alpha)N(\mathfrak{a})^{-s} \\ &= \sum_{j=0}^{4m-1} \xi(\lambda)^j \sum_{\substack{\mathfrak{a}\in\lambda^j \\ \mathfrak{a}\subset\mathfrak{o}_k}} N(\mathfrak{a})^{-s}. \end{aligned}$$

For every odd j, the correspondence

$$\mathfrak{a}\in\lambda^j,\quad \mathfrak{a}\subset\mathfrak{o}_k \Longleftrightarrow \mathfrak{a}'\in\lambda^{(2m+1)j},\quad \mathfrak{a}'\subset\mathfrak{o}_k$$

is bijective and $\xi(\lambda)^j = (-1)^j\xi(\lambda)^{(2m+1)j}$. Hence

$$\begin{aligned} L_k(s,\xi) &= \sum_{j=0}^{2m-1} \xi(\lambda^2)^j \sum_{\substack{\mathfrak{a}\in\lambda^{2j} \\ \mathfrak{a}\subset\mathfrak{o}_k}} N(\mathfrak{a})^{-s} \\ &= \sum_{j=0}^{m-1} \xi(\lambda^2)^j \{\sum_{\substack{\mathfrak{a}\in\lambda^{2j} \\ \mathfrak{a}\subset\mathfrak{o}_k}} N(\mathfrak{a})^{-s} - \sum_{\substack{\mathfrak{a}\in\lambda^{2m+2j} \\ \mathfrak{a}\subset\mathfrak{o}_k}} N(\mathfrak{a})^{-s}\}. \end{aligned}$$

Applying inverse Mollin transformation on the above L-function $L(s,\xi)$, we have the following positive definite modular form of weight 1:

$$\theta_k(\tau) = \sum_{j=0}^{m-1} \xi(\lambda^2)^j \{\theta_{2j}(\tau) - \theta_{2m+2j}(\tau)\},$$

where

$$\theta_j(\tau) = \sum_{\substack{\mathfrak{a}\in\lambda^j \\ \mathfrak{a}\subset\mathfrak{o}_k}} q^{N(\mathfrak{a})} \quad (q = e^{2\pi i\tau}).$$

It is now clear that the above results, combined with the coincidence (3.3) in Section 3.2, prove the following identity:

$$\theta_F(\tau) = \theta_k(\tau).$$

From now on, we assume again that $K/\mathbb{Q}$ is a dihedral extension. Then $m = 1$ and

$$\begin{aligned}\theta_F(\tau) &= \theta(QD_1\tau;\, 1,\, \mathfrak{o}_F,\, \mathfrak{f}) \\ &= t^{-1}\vartheta_\kappa(QD_1\tau;\, \rho,\, Q\sqrt{D}),\end{aligned}$$

where $\kappa = \pm 1$, $N(\rho)\kappa > 0$, $\mathfrak{f}\rho = (Q\sqrt{D})$ and $t = [E_{\mathfrak{f}}^{+} : \mathfrak{U}_0]$. Consequently we have

Theorem 3.1 ([44]). *The notation and assumptions being as above, we have the following identity between positive definite and indefinite cusp forms of weight* 1*:*

$$t^{-1}\vartheta_\kappa(QD_1\tau;\, \rho,\, Q\sqrt{D}) = \theta_0(\tau) - \theta_2(\tau). \tag{3.4}$$

Theorem 3.1 gives a number theoretic explanation of the identities discovered by Kac-Peterson ([58]).

3.4 Numerical examples

In this section we shall give some numerical examples based on Lemma 3.2 and Theorem 3.1 in Section 3.3. As the method for making of the examples is the same for each, we shall gives the details only for the first example.

Example 3.1. For the first example we set $F = \mathbb{Q}(\sqrt{3})$ and $\mathfrak{f} = (2\sqrt{3})$. The fundamental unit of F is totally positive and is given by $\varepsilon = 2 + \sqrt{3}$. It is easy to see that $\varepsilon^2 \equiv 1 \bmod \mathfrak{f}$. Put $\mu = (7 - 6\sqrt{3})$. Then the group $H_F(\mathfrak{f})$ is an abelian group of type $(2, 2)$:

$$H_F(\mathfrak{f}) = \{1,\, \mu,\, \mu',\, \mu\mu'\};$$

and

$$H_F(\mathfrak{f})_0 = \{1, \mu\mu'\}.$$

Hence the field F and the conductor $\mathfrak{f}$ satisfy the conditions (3.1) and (3.2) in Section 3.2. By Lemma 3.2 we know that $k = \mathbb{Q}(\sqrt{-1})$ and $\mathfrak{c} = (6)$. Furthermore, since $H_k(\mathfrak{c})$ is a group of order 4, we have $C = \{1\}$, and so

$$H_k(\mathfrak{c}) = \langle\lambda\rangle, \quad \lambda = (2 + \sqrt{-1}).$$

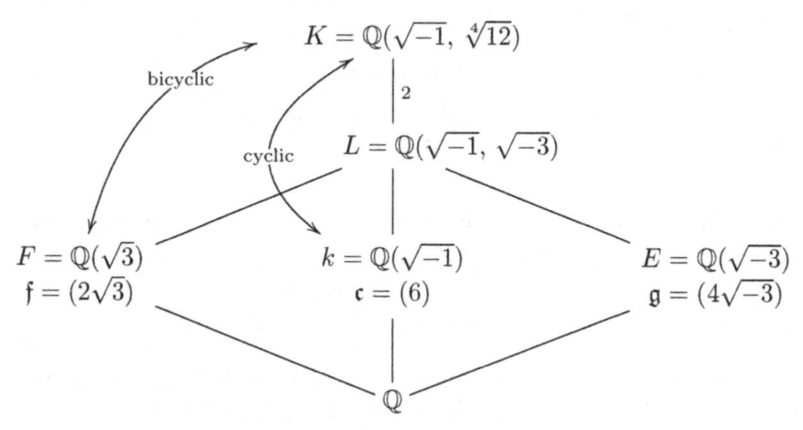

In the following we shall look for the explicit forms of θ_k and θ_F. First we treat the function $\theta_k(\tau)$. It is easy to see that

$$\begin{cases} \mathfrak{a} \in (1) \Longleftrightarrow \mathfrak{a} = (\alpha),\, \alpha \equiv 1 \pmod 6, \\ \mathfrak{a} \in \lambda^2 \Longleftrightarrow \mathfrak{a} = (\alpha),\, \alpha \equiv 2 + 3\sqrt{-1} \pmod 6. \end{cases}$$

Hence, if $\alpha = x + 3\sqrt{-1}y$ $((x,3) = 1)$, then we have

$$\begin{cases} (\alpha) \in (1) \Longleftrightarrow & x \equiv 1 \pmod 2 \text{ and } y \equiv 0 \pmod 2, \\ (\alpha) \in \lambda^2 \Longleftrightarrow & x \equiv 0 \pmod 2 \text{ and } y \equiv 1 \pmod 2. \end{cases}$$

Therefore

$$\begin{aligned} \theta_k(\tau) &= \frac{1}{2} \sum_{\substack{x,y \in \mathbb{Z} \\ (x,3)=1,\, x \not\equiv y \pmod 2}} (-1)^y q^{x^2+9y^2} \\ &= \eta^2(12\tau) \quad (q = e^{2\pi i \tau}). \end{aligned}$$

Next, for the function $\theta_F(\tau)$,

$$\begin{cases} \mathfrak{a} \in (1) \Longleftrightarrow \mathfrak{a} = (\alpha),\, \alpha > 0,\, \alpha' > 0 \text{ and } \alpha \equiv 1 \pmod{2\sqrt{3}}, \\ \mathfrak{a} \in \mu\mu' \Longleftrightarrow \mathfrak{a} = (\alpha),\, \alpha > 0,\, \alpha' > 0 \text{ and } \alpha \equiv -1 \pmod{2\sqrt{3}}. \end{cases}$$

Therefore, if $\alpha = x + 3\sqrt{3}y$ $(x \equiv \pm 1 \pmod 6)$, we have

$$\begin{cases} (\alpha) \in (1) \Longleftrightarrow x \equiv 1 \pmod 3, \\ (\alpha) \in \mu\mu' \Longleftrightarrow x \equiv -1 \pmod 3. \end{cases}$$

Since $\alpha\varepsilon^{\pm 2} = (7x \pm 24y) + (14y \pm 4x)\sqrt{3}$, we have the following as a fundamental domain:

$$x \geqq 4|y|,$$

so that we have the identities[2]

$$\theta_F(\tau) = \vartheta_+(12\tau;\, 1,\, \sqrt{12}) = \sum_{\substack{x,y\in\mathbb{Z} \\ x\geqq 4|y|,\,(x,6)=1}} \left(\frac{x}{3}\right) q^{x^2-12y^2}.$$

Another form of $\theta_F(\tau)$ is obtained as follows. Let ρ be any positive integer in F. Then it is easy to see that

$$\theta_F(\tau) = \sum_{\beta} (\operatorname{sgn}\beta) q^{N(\beta)/N(\rho)},$$

where β in the sum runs over all integers of F such that $\beta \equiv \rho \bmod \mathfrak{f}\rho$, $\beta \bmod E_{\mathfrak{f}}^{+}$ and $N(\beta)N(\rho) > 0$. Now we set $\rho = 1+\sqrt{3}$. Put

$$\beta = \begin{cases} x + y\sqrt{3}, \text{ if } \beta > 0, \\ x - y\sqrt{3}, \text{ if } \beta < 0 \end{cases}$$

for rational integers x and y. Then, for the case $\beta > 0$,

$$y > 0, \quad x \equiv 1 \pmod 6 \quad \text{and} \quad x \equiv y \pmod 4.$$

Therefore we can put

$$x = 6l+1, \quad y = 2k+1 \quad \text{with} \quad k \equiv l \pmod 2$$

for rational integers k and l. Since $\beta\varepsilon^{\pm 2} = (7x \pm 12y) + (7y \pm 4x)\sqrt{3}$, we have $7y \pm 4x \geqq y$, i.e., $3y \geqq 2|x|$; and hence $k \geqq 2|l|$. For the case $\beta < 0$, we have $y > 0$, $x \equiv 1 \pmod 6$ and $x \equiv y + 2 \pmod 4$. Hence we put

$$x = 6l+1, \quad y = 2k+1 \quad \text{with} \quad k \not\equiv l \pmod 2$$

for rational integers k and l. Since $\beta\varepsilon^{\pm 2} = (7x \mp 12y) + (-7y \pm 4x)\sqrt{3}$, we also have the following as a fundamental domain: $k \geqq 2|l|$. Therefore we obtain the following expression[3] of $\theta_F(\tau)$:

$$\theta_F(\tau) = \sum_{\substack{k,l\in\mathbb{Z} \\ k\geqq 2|l|}} (-1)^{k+l} q^{(3(2k+1)^2-(6l+1)^2)/2}.$$

For comparison, we write down the expression of the above right-hand side by Hecke's modular form:

$$\vartheta_-(12\tau;\, 1+\sqrt{3},\, (1+\sqrt{3}),\, \sqrt{12}) = \sum_{\substack{k,l\in\mathbb{Z} \\ k\geqq 2|l|}} (-1)^{k+l} q^{(3(2k+1)^2-(6l+1)^2)/2}.$$

[2] Hecke also found this expression ([34], pp. 425-426).

[3] Cf. Rogers ([79], p. 323).

By combining the above results and the identity (3.4), we have the following remarkable identities:

$$\theta_F(\tau) = \vartheta_+(12\tau;\, 1,\, \sqrt{12}) = \sum_{\substack{x,y\in\mathbb{Z} \\ x\geqq 4|y|,\, (x,6)=1}} \left(\frac{x}{3}\right) q^{x^2-12y^2}$$

$$= \sum_{\substack{k,l\in\mathbb{Z} \\ k\geqq 2|l|}} (-1)^{k+l} q^{(3(2k+1)^2-(6l+1)^2)/2}$$

$$= \theta_k(\tau) = \frac{1}{2} \sum_{\substack{x,y\in\mathbb{Z} \\ (x,3)=1,\, x\not\equiv y \pmod 2}} (-1) q^{x^2+9y^2} = \eta^2(12\tau),$$

where $\eta(\tau)$ is the *Dedekind eta function.* In exactly the same way as for $\theta_k(\tau)$, we obtain

$$\theta_E(\tau) = \sum_{k,l\in\mathbb{Z}} (-1)^{k+l} q^{(6k+1)^2+12l^2}$$

$$= \eta(24\tau)\theta_0(24\tau) \quad (= \eta^2(12\tau)),$$

where

$$\theta_0(\tau) = \sum_{m\in\mathbb{Z}} (-1)^m e^{\pi i m^2 \tau}.$$

Example 3.2. We set $F = \mathbb{Q}(\sqrt{2})$ and $\mathfrak{f} = (4)$. The fundamental unit of F is given by $\varepsilon = 1 + \sqrt{2}$ and satisfies $N(\varepsilon) = -1$ and $\varepsilon^4 \equiv 1 \bmod \mathfrak{f}$. Thus, in the same way as for the first example, we have

$$\begin{cases} k = \mathbb{Q}(\sqrt{2}), & \mathfrak{c} = (4), \\ E = \mathbb{Q}(\sqrt{-1}), & \mathfrak{g} = (4(1+\sqrt{-1})), \\ K = \mathbb{Q}(\sqrt{\varepsilon}); \end{cases}$$

and obtain the following identities:

$$\theta_F(\tau) = \vartheta_+(8\tau;\, 2+\sqrt{2},\, 2\sqrt{8})$$

$$= \sum_{\substack{x,y\in\mathbb{Z} \\ x\geqq 6|y|,\, (x,2)=1}} \left(\frac{-1}{x}\right) q^{x^2-32y^2} = \sum_{\substack{m,n\in\mathbb{Z} \\ n\geqq 3|m|}} (-1)^n q^{(2n+1)^2-32m^2}$$

$$= \theta_k(\tau) = \sum_{\substack{x,y\in\mathbb{Z} \\ x\equiv 1 \pmod 4}} (-1)^y q^{x^2+8y^2}$$

$$= \sum_{m,n\in\mathbb{Z}} (-1)^n q^{(4m+1)^2+8n^2} = \eta(8\tau)\eta(16\tau)$$

$$= \theta_E(\tau) = \sum_{m,n\in\mathbb{Z}} (-1)^{m+n} q^{(4m+1)^2+16n^2}.$$

Example 3.3.

$$\begin{cases} F = \mathbb{Q}(\sqrt{5}), \quad \mathfrak{f} = (4);\ \varepsilon = \frac{1+\sqrt{5}}{2},\ N(\varepsilon) = -1,\ \varepsilon^6 \equiv 1 \bmod \mathfrak{f}, \\ k = \mathbb{Q}(\sqrt{-5}),\ \mathfrak{c} = (2), \\ F = \mathbb{Q}(\sqrt{-1}),\ \mathfrak{g} = (10), \\ K = k(\sqrt{\varepsilon}). \end{cases}$$

$$\begin{aligned} \theta_F(\tau) &= \frac{1}{2}\vartheta_+(4\tau;\ (5+\sqrt{5})/2,\ 4\sqrt{5}) \\ &= \sum_{\substack{x,y\in\mathbb{Z} \\ x\geqq 5|y|,\ (x,2)=1}} (-1)^{y+(x-1)/2} q^{x^2-20y^2} \\ &= \sum_{\substack{k,l\in\mathbb{Z} \\ 2k\geqq l\geqq 0}} (-1)^k q^{(5(2k+1)^2-(2l+1)^2)/4} \\ &= \theta_k(\tau) = \frac{1}{2} \sum_{\substack{x,y\in\mathbb{Z} \\ x\not\equiv y\ (\mathrm{mod}\ 2)}} (-1)^y q^{x^2+5y^2}. \end{aligned}$$

The second expression of $\theta_k(\tau)$ is obtained as follows: It is clear that $H_k(\mathfrak{c})$ is a cyclic group of order 4 and

$$H_k(\mathfrak{c}) = \langle\lambda\rangle, \quad \lambda = [3, 1+\sqrt{-5}].$$

By the result in Section 3.3., we have also

$$L_k(s,\xi) = \sum_{\substack{\mathfrak{a}\in(1) \\ \mathfrak{a}\subset\mathfrak{o}_k}} N(\mathfrak{a})^{-s} \sum_{\substack{\mathfrak{a}\in\lambda^2 \\ \mathfrak{a}\subset\mathfrak{o}_k}} N(\mathfrak{a})^{-s}. \tag{3.5}$$

In the following we shall calculate the right-hand side of this equality. We can put

$$\mathfrak{a} = (\mu), \quad \mu = a + b\sqrt{-5} \quad (a, b \in \mathbb{Z}).$$

Thus

$$\begin{cases} \mathfrak{a} \in (1) \iff \mu \equiv 1\,(\mathrm{mod}\,2) \iff a \equiv 1 \quad \text{and} \quad b \equiv 0\,(\mathrm{mod}\,2), \\ \mathfrak{a} \in \lambda^2 \iff \mu \equiv 2-\sqrt{5}\,(\mathrm{mod}\,2) \iff a \equiv 0 \quad \text{and} \quad b \equiv 1\,(\mathrm{mod}\,2). \end{cases}$$

The contribution of ideals $\mathfrak{a}$ divided by λ to the first sum in (3.5) cancels that to the second sum in (3.5). Therefore we may consider the ideals $\mathfrak{a}$ with $(\mathfrak{a},\lambda) = 1$ in the above sum (3.5). Hence, if we put $\mu = (2a+1) + 2b\sqrt{-5}$ $(a, b \in \mathbb{Z})$, we have $2(a-b)+1 \not\equiv 0\,(\mathrm{mod}\,3)$. On the other hand,

$$(1-\sqrt{-5})\mu = (2a+10b+1) + (2(b-a)-1)\sqrt{-5}.$$

Put $s = b - a$ and $t = a + 5b$, then $t \equiv 5s \pmod 6$. Therefore we put $s = u + 8m$ and $t = v + 6n$. Then $v \equiv 5u \pmod 6$ $(0 \leqq u, v \leqq 5)$. Hence

$$2(b-a) \equiv 1 \pmod 3 \iff 2u - 1 \equiv 0 \pmod 3 \iff u = 2, 5.$$

Therefore

$$(u, v) = (0,0),\ (1,5),\ (3,3) \text{ and } (4,2);$$

and

$$N(\mu) = \{(12n + 2v + 1)^2 + 5(12m + 2u - 1)^2\}/6$$

Now we obtain

$$\begin{aligned}\sum_{\substack{\mathfrak{a}\in(1)\\ \mathfrak{a}\subset\mathfrak{o}_k\\ (\mathfrak{a}.\lambda)=1}} N(\mathfrak{a})^{-s} &= \frac{1}{2}\left\{\sum_{m,n\in\mathbb{Z}} 2\left(\frac{(12n+7)^2 + 5(12m+7)^2}{6}\right)^{-s}\right.\\ &\qquad \left. + \sum_{m,n\in\mathbb{Z}} 2\left(\frac{(12n+1)^2 + 5(12m+1)^2}{6}\right)^{-s}\right\}\\ &= \sum_{\substack{m,n\in\mathbb{Z}\\ m\equiv n \pmod 2}} (-1)^{m+n}\left(\frac{(6n+1)^2 + 5(6m+1)^2}{6}\right)^{-s}\end{aligned}$$

In the same way as above, we obtain

$$\sum_{\substack{\mathfrak{a}\in\lambda^2\\ \mathfrak{a}\subset\mathfrak{o}_k\\ (\mathfrak{a}.\lambda)=1}} N(\mathfrak{a})^{-s} = \sum_{\substack{m,n\in\mathbb{Z}\\ m+n\equiv 1 \pmod 2}} (-1)^{m+n}\left(\frac{(6n+1)^2 + 5(6m+1)^2}{6}\right)^{-s}.$$

Therefore we have

$$L_k(s,\xi) = \sum_{m,n\in\mathbb{Z}} (-1)^{m+n}\left(\frac{(6n+1)^2 + 5(6m+1)^2}{6}\right)^{-s}$$

Hence

$$\begin{aligned}\theta_k(\tau) &= \sum_{m,n\in\mathbb{Z}} (-1)^{m+n} q^{((6n+1)^2+5(6m+1)^2)/6}\\ &= \eta(4\tau)\eta(20\tau).\end{aligned}$$

Example 3.4.

$$\begin{aligned}&F = \mathbb{Q}(\sqrt{21}), \quad \mathfrak{f} = \left(\frac{3+\sqrt{21}}{2}\right); \quad \varepsilon = \frac{5+\sqrt{21}}{2} \equiv 1 \bmod \mathfrak{f},\\ &k = \mathbb{Q}(\sqrt{-7}), \quad \mathfrak{c} = (3),\\ &E = \mathbb{Q}(\sqrt{-3}),\\ &K = k(\sqrt{\alpha}), \quad \alpha = \frac{3+\sqrt{21}}{2}.\end{aligned}$$

$$\theta_F(\tau) = \sum_{\substack{x,y\in\mathbb{Z}\\ x\geqq 7|y|,\, x\equiv y \,(\mathrm{mod}\, 2)}} \left(\frac{-x}{3}\right) q^{(x^2-21y^2)/4} = \frac{1}{2}\vartheta_+(3\tau;\, (7+\sqrt{21})/2,\, \sqrt{21})$$
$$= \theta_k(\tau) = \sum_{\substack{x,y\in\mathbb{Z}\\ x\equiv y\,(\mathrm{mod}\, 2)}} \sigma(x,y) q^{(x^2+7y^2)/4},$$

where

$$\sigma(x,y) = \begin{cases} 1, & \text{if } 3 \mid y \text{ and } 3 \nmid x, \\ -1, & \text{if } 3 \mid x \text{ and } 3 \nmid y, \\ 0, & \text{otherwise.} \end{cases}$$

On the other hand, after a computation similar to that in Example 3, we find

$$\theta_k(\tau) = \sum_{m,n\in\mathbb{Z}} (-1)^{m+n} q^{((6m+1)^2+7(6n-1)^2)/8}$$
$$= \eta(3\tau)\eta(21\tau).$$

Remark 3.1. The indefinite representations in Example 3.1–3.3 were discovered by Kac-Peterson ([58]) by using the general theory of string functions for infinite-dimensional affine Lie algebras. A similar result was obtained for some other cases ([56]).

Remark 3.2. $\eta(\tau)\eta(23\tau)$, $\eta(2\tau)\eta(22\tau)$ and $\eta(6\tau)\eta(18\tau)$ are D_3-type and hence can not be expressed by indefinite theta series.

Remark 3.3. Biquadratic residue mod p and cusp forms of weight 1. In example 3.2, we have obtained the following identity

$$\sum_{m,n\in\mathbb{Z}} (-1)^n q^{(4m+1)^2+8n^2} = \sum_{m,n\in\mathbb{Z}} (-1)^{m+n} q^{(4m+1)^2+16n^2}, \tag{3.6}$$

by intermediating the function $\theta_F(\tau)$. This identity appeared for the first time in Jacobi's memoir and gives a generalization of the equivalence of Gauss' two criteria for the biquadratic residuacity of 2. In the following, we shall discuss more precisely this fact from our point of view. Consider the following diagram:

$$K = \mathbb{Q}(i,\varepsilon), \qquad \varepsilon = 1+\sqrt{2}$$
$$K' = \mathbb{Q}(i, \sqrt[4]{2}), \quad i = \sqrt{-1}$$

Ω

$K \qquad K'$

L

$F = \mathbb{Q}(\sqrt{2}) \quad E = \mathbb{Q}(i) \quad k = \mathbb{Q}(\sqrt{-2})$

$\mathbb{Q}$

Then, at the same time, Ω is the maximal ray class field over $F \bmod 4\sqrt{2}(\infty_1)(\infty_2)$, over $k \bmod 4\sqrt{-2}$ and over $E \bmod 8$, where ∞_i ($i = 1, 2$) denote two infinite places of F. Let p and r be distinct primes such that $p \equiv r \equiv 1 \pmod 4$. We write $\left(\frac{r}{p}\right)_4 = 1$ or -1, according as r is or is not a fourth-power residue mod p. Then it is easily checked that

$$p \text{ splits completely in } L \Longleftrightarrow \left(\frac{-1}{p}\right) = \left(\frac{-2}{p}\right) = 1$$

$$\Longleftrightarrow p \equiv 1 \pmod 8 \Longleftrightarrow p = (4a+1)^2 + 8b^2 \Longleftrightarrow p = (4\alpha+1)^2 + 16\beta^2;$$

and moreover

$$\left(\frac{\varepsilon}{p}\right) = 1 \Longleftrightarrow p \text{ splits completely in } K \qquad (3.7)$$

$$\Longleftrightarrow b \equiv 0 \pmod 2 \Longleftrightarrow \alpha + \beta \equiv 0 \pmod 2,$$

and

$$\left(\frac{2}{p}\right)_4 = 1 \Longleftrightarrow p \text{ splits completely in } K' \qquad (3.8)$$

$$\Longleftrightarrow a \equiv 0 \pmod 2 \Longleftrightarrow \beta \equiv 0 \pmod 2.$$

The above identity (3.6) gives a generalization of the equivalence (3.7); and the following identity gives a generalization of (3.8):

$$\sum_{\alpha,\beta\in\mathbb{Z}} (-1)^{\beta} q^{(4\alpha+1)^2+16\beta^2} = \sum_{a,b\in\mathbb{Z}} (-1)^{a} q^{(4a+1)^2+8b^2},$$

$$= \frac{1}{2}\theta_2(8\tau)\theta_0(32\tau),$$

where

$$\theta_2(\tau) = \sum_{m\equiv 1 \pmod 2} e^{\pi i m^2 \tau/4}.$$

We shall discuss a more general case in the second half of this chapter.

3.5 Higher reciprocity laws for some real quadratic fields

Let F be a real quadratic field satisfying the conditions (3.1) and (3.2) in Section 3.2. Then there exists an imaginary quadratic field k, and two L-functions associated with F and k are coincident. Suppose that K/k is a cyclic extension and $K/\mathbb{Q}$ a dihedral extension. Let $f(x)$ be a defining polynomial with integer coefficients of $K/\mathbb{Q}$ through the real quadratic field F. Then we have the following higher reciprocity law for $f(x)$:

Theorem 3.2. $\mathrm{Spl}\{f(x)\} = \{p\text{: prime } \mid p \nmid D_f,\, a(p) = 2\}$, *where* D_f *denotes the discriminant of* f, *and* $a(p)$ *denotes* p*-th Fourier coefficient of Hecke's indefinite modular form* $\theta_F(\tau)$ *associated with* F.

Proof. We put

$$\theta_k(\tau) = \sum_{\mathfrak{a}\subset\mathfrak{o}_k} \xi(\mathfrak{a})q^{N(\mathfrak{a})} = \sum_{n=1}^{\infty} b(n)q^n.$$

Let $\mathfrak{p}$ be any prime ideal of k unramified for K/k. Then we know that

(i) $\xi(\mathfrak{p}) = 1 \Longleftrightarrow \mathfrak{p} \in (1) \Longleftrightarrow \mathfrak{p}$ splits completely in K;
(ii) $\xi(\mathfrak{p}) = -1 \Longleftrightarrow \mathfrak{p} \in \lambda^2 \Longleftrightarrow \mathfrak{p}$ splits completely in L/k and remains prime in K/L;
(iii) $\xi(\mathfrak{p}) = i$ or $-i \Longleftrightarrow \mathfrak{p} \in \lambda$ or $\mathfrak{p} \in \lambda^3 \Longleftrightarrow \mathfrak{p}$ remains prime in K.

Let p be a prime number and $p = \mathfrak{p}\mathfrak{p}'$ in k, where $\mathfrak{p}'$ denotes the conjugate of $\mathfrak{p}$. Then

$$\mathfrak{p} \in (1) \Longrightarrow b(p) = 2;$$

and vice versa. Let $F(x)$ be a defining polynomial with integer coefficients of K/k. Then it is easy to see that

$$\mathrm{Spl}\{F(x)\} = \{p \mid p \nmid D_F,\, b(p) = 2\},$$

where D_F denotes the discriminant of F. On the other hand,

$$\begin{aligned}&\mathrm{Spl}\{f(x)\} \cup \{p \,|\, p \text{ unramified}, p \nmid D_f\}\\ &\quad = \mathrm{Spl}\{F(x)\} \cup \{p \,|\, p \text{ unramified}, p \nmid D_F\};\end{aligned}$$

and by Theorem 3.1, $b(p) = a(p)$ for all p. Hence we obtain

$$\mathrm{Spl}\{f(x)\} = \{p \mid p \nmid D_f,\, a(p) = 2\}.$$ □

Example 3.5. We shall use the same symbols as in Example 3.1. Then we have the following defining equation of K/k:

$$F(x) = x^4 - 6x^2 - 3.$$

On the other hand a defining equation of K/F is given by

$$f_1(x) = x^4 - 4(1+\sqrt{3})x^2 + 4(2+\sqrt{3})^2.$$

Therefore the following is a defining equation of $K/\mathbb{Q}$ through the field F:

$$\begin{aligned}f(x) &= f_1(x)\cdot f_1(x)'\\ &= x^8 - 8x^6 + 24x^4 + 160x^2 + 16.\end{aligned}$$

Hence

$$\begin{aligned}\mathrm{Spl}\{F(x)\} &= \mathrm{Spl}\{f(x)\} = \{p \mid a(p) = 2\} \\ &= \{p \mid p = u^2 + v^2,\ u \equiv 0 \pmod 6),\ u, v \in \mathbb{Z}\},\end{aligned}$$

where

$$\theta_F(\tau) = \vartheta_+(12\tau;\ 1,\ \sqrt{12}) = \sum_{n=1}^{\infty} a(n)q^n.$$

Remark 3.4. For the defining polynomial $f(x)$ in Theorem 3.2, the following assertions hold:

1) $f(x) \bmod p$ has exactly 2 distinct quartic factors over F_p
$\Longleftrightarrow a(p) = 0$ and $a(p^2) = -1$;

2) $f(x) \bmod p$ has exactly 4 distinct quadratic factors over F_p
$\Longleftrightarrow$ '$a(p) = -2$' or '$a(p) = 0$ and $a(p^2) = 1$'.

3.6 Cusp forms of weight 1 related to quartic residuacity

Let m be a positive square-free integer and ε_m be the fundamental unit of the real quadratic field $\mathbb{Q}(\sqrt{m})$. We consider only those m for which ε_m has norm $+1$. Let K be the Galois extension of degree 16 over $\mathbb{Q}$ generated by $\sqrt{-1}$ and $\sqrt[4]{\varepsilon_m}$ and we put $G = \mathrm{Gal}\,(K/\mathbb{Q})$. Then the group G is generated by three elements σ, ϕ and ρ in such way that

$$\begin{aligned}\sigma(\sqrt[4]{\varepsilon_m}) &= \sqrt{-1}\sqrt[4]{\varepsilon_m}, \\ \phi(\sqrt[4]{\varepsilon_m}) &= \sqrt[4]{\varepsilon_m}^{-1}, \\ \rho(\sqrt{-1}) &= -\sqrt{-1},\end{aligned}$$

and has defining relations:

$$\sigma^4 = \phi^2 = \rho^2 = 1, \quad \phi\rho = \rho\phi, \quad \rho\sigma\rho = \phi\sigma\phi = \sigma^3.$$

The group G has three abelian subgroups of index 2 in G, which are the following:

$$\begin{aligned}H_k &= \langle \sigma, \phi\rho \rangle & &\Longleftrightarrow k = \mathbb{Q}(\sqrt{-m}), \\ H_F &= \langle \sigma^2, \phi, \rho \rangle & &\Longleftrightarrow F = \mathbb{Q}(\sqrt{t+2}), \\ H_E &= \langle \sigma^2, \sigma\phi, \sigma\rho \rangle & &\Longleftrightarrow E = \mathbb{Q}(\sqrt{-m(t+2)}),\end{aligned}$$

where $t = \mathrm{tr}(\varepsilon_m)$. Let f and e be the square-free part of $t+2$ and $m(t+2)$, respectively, and put

$$K' = \mathbb{Q}(\sqrt{-1}, \sqrt{\varepsilon_m}), \quad L = \mathbb{Q}(\sqrt{-1}, \sqrt{-m}),$$

$$L' = \mathbb{Q}(\sqrt{-m}, \sqrt{f}), \quad L'' = \mathbb{Q}(\sqrt{-m}, \sqrt{-f}).$$

Then we have the following diagram:

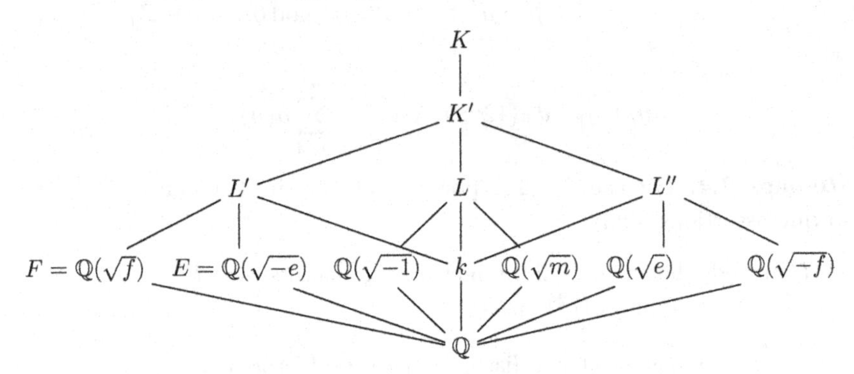

By this diagram, we have the following equivalence for any odd prime l:

$$l \text{ splits completely in } K' \Longleftrightarrow \left(\frac{-1}{l}\right) = \left(\frac{f}{l}\right) = \left(\frac{e}{l}\right) = 1, \tag{3.9}$$

where $\left(\frac{*}{l}\right)$ denotes the Legendre symbol. The group G has the following eight representations Γ_i of degree 1, where $j = 1, \ldots, 8$.

	Γ_1	Γ_2	Γ_3	Γ_4	Γ_5	Γ_6	Γ_7	Γ_8
σ	1	1	1	1	−1	−1	−1	−1
ϕ	1	1	−1	−1	1	1	−1	−1
ρ	1	−1	1	−1	1	−1	1	−1

The group G has just two irreducible representations of degree 2, which have determinant Γ_4. If we denote by ψ_0 the one of these, then the other is $\psi_0 \otimes \Gamma_3$. Let σ_l denote the Frobenius substitution associated with l in K. Then we have the following table which gives the correspondence between quadratic subfields of K and Γ_i $(2 \leqq j \leqq 8)$.

	Γ_2	Γ_3	Γ_4	Γ_5	Γ_6	Γ_7	Γ_8
	$\mathbb{Q}(\sqrt{-1})$	$\mathbb{Q}(\sqrt{m})$	k	F	$\mathbb{Q}(\sqrt{-1})$	$\mathbb{Q}(\sqrt{e})$	E
$\Gamma_j(\sigma_l)$	$\left(\frac{-1}{l}\right)$	$\left(\frac{m}{l}\right)$	$\left(\frac{-m}{l}\right)$	$\left(\frac{f}{l}\right)$	$\left(\frac{-f}{l}\right)$	$\left(\frac{e}{l}\right)$	$\left(\frac{-e}{l}\right)$

Put $\psi_1 = \psi_0 \otimes \Gamma_3$ and let $L(s, \psi_0, K/\mathbb{Q})$ (resp. $L(s, \psi_1, K/\mathbb{Q})$) denote the Artin L-function associated with ψ_0 (resp. ψ_1) and let $\Theta(\tau; \psi_0)$

(resp. $\Theta(\tau;\ \psi_1)$) denote the Mellin transformation of $L(s, \psi_0, K/\mathbb{Q})$ (resp. $L(s, \psi_1, K/\mathbb{Q})$). Then we can define the following function:

$$\Theta(\tau;\ K) = \frac{1}{2}\{\Theta(\tau;\ \psi_0) + \Theta(\tau;\ \psi_1)\}.$$

Let N denote the L.C.M of the conductor of ψ_0 and that of ψ_1. Then the function $\Theta(\tau;\ K)$ is a cusp form of weight 1 on the congruence subgroup $\Gamma_0(N)$ with the character $\left(\frac{-m}{l}\right)$.

Let M be one of the three quadratic fields k, E and F. Then K is abelian over M. Let $\mathfrak{o}_M$ be the ring of integers of M and $\mathfrak{a}$ an ideal of $\mathfrak{o}_M$. If M is imaginary (resp. real), then $H_M(\mathfrak{a})$ denotes the group of ray classes (resp. narrow ray classes) modulo $\mathfrak{a}$ of M. Let $\mathfrak{b}$ be an ideal of M prime to $\mathfrak{a}$ and $[\mathfrak{b}]$ the class in $H_M(\mathfrak{a})$ represented by $\mathfrak{b}$. If in particular b is an element of M, then the ideal class $[(b)]$ represented by the principal ideal (b) is abbreviated as $[b]$. Let $\bar{\mathfrak{f}}(K/M)$ (resp. $\mathfrak{f}(K/M)$) be the conductor (resp. the finite part of conductor) of K over M. Furthermore we denote by $C_M(K)$ (resp. $C_M(K')$) the subgroup of $H_M(\mathfrak{f}(K/M))$ corresponding to K (resp .K'). The restriction ψ_0 (resp. ψ_1) to the abelian Galois group $G(K/M)$ decomposes into distinct linear representations ξ_M and ξ'_M (resp. $\xi_M \otimes \Gamma_3$ and $\xi'_M \otimes \Gamma_3$) of $G(K/M)$:

$$\psi_i \,|\, G(K/M) = \xi_M \otimes \Gamma_3^i + \xi'_M \otimes \Gamma_3^i, \quad (i = 0, 1).$$

By Artin reciprocity law, we can identity ξ_M and ξ'_M with characters of $H_M(\mathfrak{f}(k/M))$ trivial on $C_M(K)$ and so we denote these characters by the same notation. Let c_M be the finite part of conductor of ξ_M. We assume that the finite part of conductor of $\xi_M \otimes \Gamma_3$ is equal to c_M. Let $\widetilde{C_M(K)}$ (resp. $\widetilde{C_M(K')}$) be the image of $C_M(K)$ (resp. $C_M(K')$) by the canonical homomorphism of $H_M(\mathfrak{f}(K/M))$ to $H_M(c_M)$. Since K is the class field over M with conductor $\mathfrak{f}(K/M)$, the Artin L-function $L(s, \psi_0, K/\mathbb{Q})$ (resp. $L(s, \psi_1, K/\mathbb{Q})$) is coincident with the L-function $L_M(s, \tilde{\xi}_M)$ (resp. $L_M(s, \widetilde{\xi_M \otimes \Gamma_3})$) of M associated with the character $\tilde{\xi}_M$ (resp. $\widetilde{\xi_M \otimes \Gamma_3}$), where $\tilde{\xi}_M$ (resp. $\widetilde{\xi_M \otimes \Gamma_3}$) denotes the primitive character corresponding to ξ_M (resp. $\xi_M \otimes \Gamma_3$). Then we shall have three expressions of $\Theta(\tau;\ K)$ (Theorem 3.3 below).

Proposition 3.1. *The notation and the assumption being as above, we have*

$$\Theta(\tau;\ K) = \sum_{\substack{\mathfrak{a} \subset \mathfrak{o}_M \\ [\mathfrak{a}] \in \widetilde{C_M(K')}}} \chi_M(\mathfrak{a}) q^{N_{M/\mathbb{Q}}(\mathfrak{a})} \qquad (q = e^{2\pi i \tau}), \tag{3.10}$$

where

$$\chi_M(\mathfrak{a}) = \begin{cases} 1, \; if \; [\mathfrak{a}] \in \widetilde{C_M(K)}, \\ -1, \; otherwise; \end{cases}$$

and $N_{M/\mathbb{Q}}(\mathfrak{a})$ *denotes the norm of* $\mathfrak{a}$ *with respect to* $M/\mathbb{Q}$.

The proof of Proposition 3.1 is quite similar to that appeared in Section 3.3.

Let $f(x)$ be a defining polynomial of $\sqrt[4]{\varepsilon_m}$ over $\mathbb{Q}$. Then it is easy to see that

$$f(x) = (x^4 - \varepsilon_m)(x^4 - \varepsilon_m^{-1}) = x^8 - tx^4 + 1.$$

Let $a(n)$ be the n-th Fourier coefficient of the expression

$$\Theta(\tau; K) = \sum_{n=1}^{\infty} a(n)q^n.$$

Then we have the following relation:

Proposition 3.2. *Let* p *be any prime not dividing the discriminant* D_f *of* $f(x)$. *Then we have*

$$\#\{x \in \mathbb{F}_p \mid f(x) = 0\} = 1 + \left(\frac{m}{p}\right) + \left(\frac{f}{p}\right) + \left(\frac{e}{p}\right) + 2a(p). \qquad (3.11)$$

Proof. Let H be the group generated by ρ, say $H = \langle \rho \rangle$. Then H is the subgroup of G corresponding to $\mathbb{Q}(\sqrt[4]{\varepsilon_m})$. We denote by 1_H^G the character of G induced by the identity character of H. Then we have the following scalar product formulas:

$$(1_H^G | \Gamma_i) = \begin{cases} 1, \text{ if } i = 1, 3, 5, 7, \\ 0, \text{ otherwise}; \end{cases}$$

$$(1_H^G | \chi_i) = 1 \quad (i = 0, 1),$$

where χ_0 (resp. χ_1) denotes the character of ψ_0 (resp. ψ_1). Therefore, we have

$$\begin{aligned} 1_H^G(\sigma_p) &= \sum_{\substack{1 \leqq i \leqq 7 \\ i:\,\text{odd}}} \Gamma_i(\sigma_p) + \chi_0(\sigma_p) + \chi_1(\sigma_p) \\ &= 1 + \left(\frac{m}{p}\right) + \left(\frac{f}{p}\right) + \left(\frac{e}{p}\right) + 2a(p). \end{aligned}$$

On the other hand, it is easy to see that the left hand side of (3.11) is equal to $1_H^G(\sigma_p)$. This proves our proposition. □

By Propositions 3.1 and 3.2 we have the following

Corollary. $\mathrm{Spl}\{f(x)\} = \{p \mid p \nmid D_f, \; a(p) = 2\}$.

3.7 Fundamental lemmas

In this section, we shall determine the conductors $\mathfrak{f}(K/M)$, $\mathfrak{f}(K'/M)$, $\mathfrak{f}(L'/M)$ and $\mathfrak{f}(L/M)$. Let $\mathfrak{K}$, $\mathfrak{L}$ and $\mathfrak{F}$ be fields such that $\mathfrak{K} \supset \mathfrak{L} \supset \mathfrak{F}$ and $[\mathfrak{L} : \mathfrak{F}] = 2$. Assume that $\mathfrak{K}$ is abelian over $\mathfrak{F}$. We denote by $\mathfrak{d}(\mathfrak{L}/\mathfrak{F})$ the different of $\mathfrak{L}$ over $\mathfrak{F}$. For a prime ideal $\mathfrak{g}$ of $\mathfrak{L}$, let $f(\mathfrak{g})$ (resp. $g(\mathfrak{g})$) denote the $\mathfrak{g}$-exponent of $\mathfrak{f}(\mathfrak{K}/\mathfrak{L})$ (resp. $\mathfrak{d}(\mathfrak{L}/\mathfrak{F})$) and put

$$e(\mathfrak{g}) = \max\{0, g(\mathfrak{g}) - f(\mathfrak{g})\}.$$

Then we have the following

Lemma 3.3. $\mathfrak{f}(\mathfrak{K}/\mathfrak{F}) = \mathfrak{f}(\mathfrak{K}/\mathfrak{L})\mathfrak{d}(\mathfrak{L}/\mathfrak{F}) \prod_{\mathfrak{g}} \mathfrak{g}^{e(\mathfrak{g})}$.

We assume that $\mathfrak{L}$ is a Galois extension over $\mathbb{Q}$. Let $\mathfrak{o}_{\mathfrak{L}}$ be the ring of integers of $\mathfrak{L}$ and let $\mathfrak{p}$ be a prime ideal of $\mathfrak{o}_{\mathfrak{L}}$ dividing 2. We denote by $e_{\mathfrak{L}}$ the ramification exponent of $\mathfrak{p}$. Let $\mathfrak{o}_{\mathfrak{p}}$ denote the completion of $\mathfrak{o}_{\mathfrak{L}}$ with respect to $\mathfrak{p}$ and $\Pi_{\mathfrak{p}}$ a prime element of $\mathfrak{o}_{\mathfrak{p}}$. Furthermore, for $\xi \in \mathfrak{o}_{\mathfrak{p}}^{\times}$, we put

$$S_{\mathfrak{p}} = \max\{t \in Z^{+} \mid \xi \equiv \text{square mod } \Pi_{\mathfrak{p}}^{t}\}.$$

Then we have

Lemma 3.4. *If $S_{\mathfrak{p}}(\xi) < 2e_{\mathfrak{L}}$, then there exists a unique odd integer $t(< 2e_{\mathfrak{L}})$ such that*

$$\xi = \eta^2 + \delta \Pi_{\mathfrak{p}}^{t} \quad (\eta, \delta \in \mathfrak{o}_{\mathfrak{p}}^{\times});$$

and this uniquely determined t is equal to $S_{\mathfrak{p}}(\xi)$.

Lemma 3.5. *Put*

$$t_{\mathfrak{p}}(\xi) = \min\{n \in \mathbb{Z} \mid \xi \Pi_{\mathfrak{p}}^{2n} \equiv \textit{square} \bmod \Pi_{\mathfrak{p}}^{2e_{\mathfrak{L}}}, 0 \leqq n \leqq e_{\mathfrak{L}}\}.$$

If $S_{\mathfrak{p}}(\xi) < 2e_{\mathfrak{L}}$, then we have

$$S_{\mathfrak{p}}(\xi) = 2e_{\mathfrak{L}} + 1 - 2t_{\mathfrak{p}}(\xi).$$

Let α be an element of $\mathfrak{o}_{\mathfrak{L}}$ such that (α) is a square-free ideal with $((\alpha), 2) = 1$ and put $\mathfrak{K} = \mathfrak{L}(\sqrt{\alpha})$. We assume that $\mathfrak{K}$ is a Galois extension over $\mathbb{Q}$. Then $S_{\mathfrak{p}}(\alpha)$ is independent of $\mathfrak{p}$ chosen. Since $\mathfrak{K}$ and $\mathfrak{L}$ are the Galois extension over $\mathbb{Q}$, the $\mathfrak{p}$-exponent $f(\mathfrak{p})$ of $\mathfrak{f}(\mathfrak{K}/\mathfrak{L})$ does not depend on $\mathfrak{p}$ chosen. Thus we can put $S_{\mathfrak{L}}(\alpha) = S_{\mathfrak{p}}(\alpha)$ and $f(2) = f(\mathfrak{p})$.

Lemma 3.6. (i) *The prime ideal $\mathfrak{p}$ is ramified for $\mathfrak{K}/\mathfrak{L}$ if and only if $S_{\mathfrak{p}}(\alpha) < 2e_{\mathfrak{L}}$.*

(ii) *If $S_{\mathfrak{L}}(\alpha) < 2e_{\mathfrak{L}}$, then $S_{\mathfrak{L}}(\alpha)$ is equal to the odd number $t(< 2e_{\mathfrak{L}})$ determined by*

$$\alpha = \eta^2 + \delta \Pi_{\mathfrak{p}}^t \quad (\eta, \delta \in \mathfrak{o}_{\mathfrak{p}}^{\times});$$

and moreover

$$f(2) = 2e_{\mathfrak{L}} + 1 - S_{\mathfrak{L}}(\alpha).$$

Proof. By the assumption on α, we have

$$\mathfrak{o}_{\mathfrak{K}} = \left\{ \frac{1}{2}(a + b\sqrt{\alpha}) \mid a, b \in \mathfrak{o}_{\mathfrak{L}},\, a^2 - \alpha b^2 \equiv 0 \,(\mathrm{mod}\, 4) \right\}.$$

Denote by $\mathfrak{P}$ a prime ideal of $\mathfrak{K}$ dividing $\mathfrak{p}$. Let $\mathfrak{a}$ be an ideal of $\mathfrak{K}$ and denote by $\mathfrak{M}_{\mathfrak{P}}(\mathfrak{a})$ the $\mathfrak{P}$-exponent of $\mathfrak{a}$, and let ε be a generator of $G(\mathfrak{K}/\mathfrak{L})$. Then, by the definition of $f(\mathfrak{p})$,

$$f(2) = \min_{\xi \in \mathfrak{o}_k} \mathfrak{M}_{\mathfrak{p}}(\xi - \xi^{\varepsilon}). \tag{3.12}$$

Denote by X (resp. $X_{\mathfrak{p}}$) the group of all elements b of $\mathfrak{o}_{\mathfrak{L}}$ satisfying the condition

$$\alpha b^2 \equiv \text{square} \ \mathrm{mod}\, 4 \quad (\text{resp.} \ \mathrm{mod}\, \mathfrak{p}^{2e_{\mathfrak{L}}}).$$

Let $\mathfrak{M}_{\mathfrak{p}}(b)$ denote the $\mathfrak{p}$-exponent of (b). Then by (3.12), we have

$$f(2) = 2 \min_{b \in X} \mathfrak{M}_{\mathfrak{p}}(b) = 2 \min_{b \in X_{\mathfrak{p}}} \mathfrak{M}_{\mathfrak{p}}(b).$$

Therefore,

$$\begin{aligned} &\mathfrak{p} \text{ is unramified for } \mathfrak{K}/\mathfrak{L} \iff f(2) = 0 \\ &\iff \alpha \text{ is square } \mathrm{mod}\, \mathfrak{p}^{2e} \iff S_{\mathfrak{L}}(\alpha) \geqq 2e_{\mathfrak{L}}. \end{aligned}$$

If $\mathfrak{p}$ is ramified for $\mathfrak{K}/\mathfrak{L}$, then

$$\min_{b \in X_{\mathfrak{p}}} \mathfrak{M}_{\mathfrak{p}}(b) = t_{\mathfrak{p}}(\alpha).$$

By Lemma 3.5, $S_{\mathfrak{L}}(\alpha) = 2e_{\mathfrak{L}} + 1 - f(2)$. Hence by Lemma 3.4 the assertion (ii) is proved. □

Now we assume that $\mathfrak{L}(\sqrt[4]{\alpha})$ is a Galois extension over $\mathbb{Q}$. It is easy to see that there exists a subgroup R of $\mathfrak{o}_{\mathfrak{p}}^{\times}$ with order $\#(\mathfrak{o}_{\mathfrak{L}}/\mathfrak{p}) - 1$ such that $R^* = R \cup \{0\}$ is a complete system of coset representatives of $\mathfrak{o}_{\mathfrak{L}} \,\mathrm{mod}\, \mathfrak{p}$. Put

$$t = \min\{2e_{\mathfrak{L}},\, S_{\mathfrak{L}}(\alpha)\} \text{ and } u = [(t+1)/2].$$

Then there exists elements $a_0, a_1, \ldots, a_{u-1}$ of R^* such that

$$\alpha \equiv \left(a_0 + a_1 \Pi_{\mathfrak{p}} + \cdots + a_{u-1} \Pi_{\mathfrak{p}}^{u-1}\right)^2 \mathrm{mod}\, \Pi_{\mathfrak{p}}^t.$$

Lemma 3.7. (i) *If $\mathfrak{p}$ is unramified for $\mathfrak{R}/\mathfrak{L}$ and there exists a nonzero element in $\{a_i \mid i$: odd$\}$, then*

$$S_{\mathfrak{R}}(\sqrt{\alpha}) = \min\{i\text{: } odd\ |a_i \neq 0\}.$$

(ii) *If $\mathfrak{p}$ is unramified for $\mathfrak{R}/\mathfrak{L}$ and there exists a prime element Π_p of $\mathfrak{o}_\mathfrak{p}$ such that $\Pi_\mathfrak{p} \equiv \Pi_\mathfrak{P}^2 \bmod \Pi_\mathfrak{P}^{t+1}$, then*

$$S_{\mathfrak{R}}(\sqrt{\alpha}) = S_{\mathfrak{L}}(\alpha).$$

Now we put

$$\mathfrak{L} = L \text{ or } K', \qquad \text{and} \qquad \alpha = \varepsilon_m.$$

Form now on we assume that m is prime number p with $p \equiv 3 \pmod 4$. We put $\varepsilon_p = \varepsilon = A + B\sqrt{p}$. Then it is easy to verify that A is an even number. Since $A^2 - pB^2 = 1$, we have $(A+1)(A-1) = pB^2$. Therefore we can put

$$A - 1 = r^2 u,$$
$$A + 1 = s^2 v,$$

with $(ru, su) = 1$, $rs = B$ and $uv = p$ $(r, s, u, v \in \mathbb{Z}^+)$. Hence, $2 = s^2v - r^2u$. By considering this relation mod 8, we have

$$(u, v) = \begin{cases} (1, p) & \text{if } p \equiv 3 \pmod 8, \\ (p, 1) & \text{if } p \equiv 7 \pmod 8. \end{cases}$$

Since $t = \mathrm{tr}(\varepsilon) = 2A$, we have $t + 2 = 2s^2v$. Hence

$$(f, e) = \begin{cases} (2p, 2) & \text{if } p \equiv 3 \pmod 8, \\ (2, 2p) & \text{if } p \equiv 7 \pmod 8. \end{cases}$$

Therefore we have the following lemma

Lemma 3.8. *With F and E as in Section 3.6, we have*

$$(F, E) = \begin{cases} (\mathbb{Q}(\sqrt{2p}), \mathbb{Q}(\sqrt{-2})), & \textit{if } p \equiv 3 \pmod 8, \\ (\mathbb{Q}(\sqrt{2}), \mathbb{Q}(\sqrt{-2p})), & \textit{if } p \equiv 7 \pmod 8. \end{cases}$$

Now we shall calculate the conductors $\mathfrak{f}(K/M)$, $\mathfrak{f}(K'/M)$, $\mathfrak{f}(L/M)$ and $\mathfrak{f}(L'/M)$. Because the method of calculation is very similar for each of the three cases, we shall give the details only for the case of $M = k$. If we put $\mathfrak{L} = L$, then $K' = L(\sqrt{\varepsilon})$. We can take $e_L = 2$ and $\Pi_p = 1 - \sqrt{p}$. Therefore, $\varepsilon \equiv 1 - \Pi_p \pmod 2$. By Lemma 3.6, $S_L(\varepsilon) = 1$ and hence $S_{K'}(\sqrt{\varepsilon}) = 1$ by (ii) of Lemma 3.7. Therefore, again by Lemma 3.6, we have $f_{K'}(2) = 5 - 1 = 4$. Since prime factors of 2 are only ramified for

K'/L, we have $\mathfrak{f}(K'/L) = (4)$, and hence $\mathfrak{d}(K'/L) = (2)$. By $e_{K'} = 4$, $f_K(2) = 9 - 1 = 8$. Therefore $\mathfrak{f}(K/K') = (4)$. Consequently, by Lemma 1, we have

$$\begin{aligned}\mathfrak{f}(K/L) &= \mathfrak{f}(K/K')\mathfrak{d}(K'/L)\\ &= (4)(2) = (8).\end{aligned}$$

Thus we obtain the following:

$$\begin{cases}\mathfrak{f}(K/k) = \mathfrak{f}(K/L)\mathfrak{d}(L/k) = (16),\\ \mathfrak{f}(K'/k) = \mathfrak{f}(K'/L)\mathfrak{d}(L/k) = (8),\\ \mathfrak{f}(L/k) = \mathfrak{d}(L/k)^2 = (4).\end{cases}$$

Therefore our required conductors are as follows.

M		$\tilde{\mathfrak{f}}(K/M)$	$\tilde{\mathfrak{f}}(K'/M)$	$\tilde{\mathfrak{f}}(L'/M)$	$\tilde{\mathfrak{f}}(L/M)$	c_M
k		16	8	8	4	16
F	$p \equiv 3 \pmod 8$	$4\mathfrak{p}_2\infty_1\infty_2$	$(2)\infty_1\infty_2$	$\infty_1\infty_2$		$4\mathfrak{p}_2$
	$p \equiv 7 \pmod 8$	$(4\sqrt{2}p)\infty_1\infty_2$	$(2p)\infty_1\infty_2$	$(p)\infty_1\infty_2$		$4\mathfrak{p}$
E	$p \equiv 3 \pmod 8$	$4\sqrt{-2}p$	$2p$	p		$4\mathfrak{p}$
	$p \equiv 7 \pmod 8$	$4\mathfrak{p}_2$	2	1		$4\mathfrak{p}_2$

In the above table, $\mathfrak{p}$ denotes a prime ideal of M dividing p, and $\mathfrak{p}_2$ denotes a prime ideal of M dividing 2. Further ∞_i $(i = 1, 2)$ denote two infinite places of F.

3.8 Three expressions of $\Theta(\tau; K)$

For an integral ideal $\mathfrak{a}$ of M, if M is imaginary (resp. real), then $P_M(\mathfrak{a})$ denotes the subgroup of $H_M(\mathfrak{a})$ generated by principal classes (resp. principal classes represented by totally positive elements). We write simply H_M and P_M in place of $H_M(\mathfrak{f}(K/M))$ and $P_M(\mathfrak{f}(K/M))$ respectively. Suppose that $\mathfrak{a}$ divides $\mathfrak{f}(K/M)$. Then we denotes by $K(\mathfrak{a})$ the kernel of the canonical homomorphism : $P_M \to P_M(\mathfrak{a})$. Moreover we put $C_M(\ )^* = P_M \cap C_M(\ )$. In the following, we shall obtain $C_M(K)$ and $C_M(K')$ under the assumption $p \equiv 7 \pmod 8$.

***Case* 1.** $M = k = (= \mathbb{Q}(\sqrt{-p}))$.

By assumption, we have $2 = \mathfrak{p}_2\bar{\mathfrak{p}}_2$, where $\bar{\mathfrak{p}}_2$ denotes the conjugate of $\mathfrak{p}_2$. Take the two elements μ and ν of $\mathfrak{o}_k$ such that

$$\begin{cases} \mu \equiv 5 \bmod \mathfrak{p}_2^4, \\ \mu \equiv 1 \bmod \bar{\mathfrak{p}}_2^4, \end{cases} \quad \begin{cases} \nu \equiv -1 \bmod \mathfrak{p}_2^4, \\ \nu \equiv 1 \bmod \bar{\mathfrak{p}}_2^4. \end{cases}$$

Then we have the following relations: $[\mu][\bar{\mu}] = [5]$, $[\mu]^4 = [\bar{\mu}]^4 = 1$, $[\nu] = [\bar{\nu}]$ and $[\nu]^2 = 1$. We also have

$$P_k = \langle [\mu],\ [\bar{\mu}],\ [\nu] \rangle, \quad K((4)) = \langle [\mu],\ [\bar{\mu}] \rangle,$$
$$K((8)) = \langle [\mu]^2,\ [\bar{\mu}]^2 \rangle.$$

By the above table, we see that

$$\begin{aligned}[P_k : C_k(L)^*] &= [C_k(L)^* : C_k(K')^*] \\ &= [C_k(K')^* : C_k(K)^*] = 2.\end{aligned}$$

Furthermore,

$$C_k(L)^* \supset K((4)), \quad C_k(K')^* \supset K((8)),\ \not\supset K((4)),$$
$$C_k(K)^* \not\supset K((8)).$$

Hence we have

$$\begin{aligned} C_k(L)^* &= \langle [\mu], [\bar{\mu}] \rangle, \\ C_k(K')^* &= \langle [\mu]^2, [\bar{\mu}]^2, [\mu][\bar{\mu}] \rangle, \\ C_k(K)^* &\not\ni [\mu]^2, [\bar{\mu}]^2. \end{aligned}$$

Since $G(K/\mathbb{Q})$ is non-abelian and $G(K/k) \cong P_k/C_k(K)^*$, we see $[\mu]^{-1}[\mu] \notin C_k(K)^*$. Therefore, $[\mu][\bar{\mu}] \in C_k(K)^*$. Hence we have

$$C_k(K)^* = \langle [\mu][\bar{\mu}] \rangle = \langle [5] \rangle.$$

We put

$$H_k = \sum_{\mathfrak{b} \in S} [\mathfrak{b}] p_k,$$

where S denotes the index set of integral ideals $\mathfrak{b}$. Then

$$\begin{aligned} C_k(K') &= C_k(K) + C_k(K)[\mu]^2, \\ C_k(K) &= \sum_{\mathfrak{b} \in S} [\mathfrak{b}]^{-4} C_k(K)^*. \end{aligned}$$

Put $\omega = (1 + \sqrt{-p})/2$ and let $\mathfrak{a}$ be an ideal of $\mathfrak{o}_k$ with $(\mathfrak{a}, (2)) = 1$. Then, by the above relations, we have $[\mathfrak{a}] \in C_k(K')$ if and only if there exists

$\mathfrak{b} \in S$ and $\eta = x + y\omega \in \mathfrak{b}^4$ such that $x \equiv 1 \pmod 2$, $y \equiv 0 \pmod 8$ and $\mathfrak{a} = \mathfrak{b}^{-4}(\eta)$. Moreover

$$[\mathfrak{a}] \in C_k(K) \Longleftrightarrow y \equiv 0 \pmod{16}.$$

Therefore, if $M = k$, then the right hand side of (3.10) is as follows:

$$\theta(\tau; K) = \sum_{\mathfrak{b}\in S} \sum_{4x+1+4y\sqrt{-p}\in\mathfrak{b}^4} (-1)^y q^{\{(4x+1)^2+16py^2\}/N_{k/\mathbb{Q}}(\mathfrak{b})^4}. \tag{3.13}$$

Case 2. $M = F = (= \mathbb{Q}(\sqrt{2}))$.

Let α be an element of $\mathfrak{o}_F$. Then there exists an element α^* of $\mathfrak{o}_F$ such that

$$\begin{cases} \alpha^* \text{ is totally positive,} \\ \alpha^* \equiv \alpha \bmod 4\sqrt{2}, \\ \alpha^* \equiv 1 \bmod p. \end{cases}$$

Let $p = \mathfrak{p}\bar{\mathfrak{p}}$ in F, and $r(\mathfrak{p})$ denotes a generator of the multiplicative group $(\mathfrak{o}_F/\mathfrak{p})^\times$. Take a totally positive element λ of $\mathfrak{o}_F$ such that

$$\begin{cases} \lambda \equiv 1 \bmod 4\sqrt{2}, \\ \lambda \equiv r(\mathfrak{p}) \bmod \mathfrak{p}, \\ \lambda \equiv 1 \bmod \bar{\mathfrak{p}}. \end{cases}$$

Then we obtain

$$C_F(L') = \langle [\varepsilon_2^*], [3^*], [5^*], [\lambda], [\bar{\lambda}], [\lambda]^2 \rangle.$$

Since the Galois group $G(K'/\mathbb{Q})$ is isomorphic to $P_F/C_F(K')$, we have

$$C_F(K') \ni [\lambda]^2, [\bar{\lambda}]^2, [\lambda]^{-1}[\bar{\lambda}].$$

Hence

$$C_F(K') = \langle [\varepsilon_2^*]^2, [3^*], [5^*], [\lambda]^2, [\bar{\lambda}]^2, [\lambda][\bar{\lambda}] \rangle.$$

Next we shall calculate $C_F(K)$. First we notice that

$$\begin{cases} C_F(K) \ni [\lambda]^2, [\bar{\lambda}]^2, [\varepsilon_2^*]^2, \\ C_F(K) \not\ni [5^*]. \end{cases}$$

Take a prime q such that $q \equiv 3 \bmod 8$ and $(q/p) = -1$. Then q remains prime in F and $[q] = [3^*]([\lambda][\bar{\lambda}])^a$ (a: odd). Since $(-p/q) = -1$, q remains prime in k also. Hence by the result of Case 1, q splits completely for K/k. Therefore, $[q] \in C_F(K)$, i.e.,

$$C_F(K) \ni [3^*]([\lambda][\bar{\lambda}]).$$

Similarly, $[5^*]([\lambda][\bar{\lambda}]) \in C_F(K)$. Therefore we obtain

$$C_F(K) = \langle [\varepsilon_2^*]^2,\, [\lambda]^2,\, [\bar{\lambda}]^2,\, [3^*][\lambda][\bar{\lambda}],\, [5^*][\lambda][\bar{\lambda}] \rangle,$$
$$C_F(K') = C_F(K) + C_F(K)[5^*].$$

Let r be a rational integer with $r^2 \equiv 2 \pmod{p}$ and $\mu = x + y\sqrt{2}$ be a totally positive element of $\mathfrak{o}_F$ such that $(2p, \mu) = 1$. Then we have

$$[\mu] \in C_F(K') \Longleftrightarrow x\text{: odd, } y\text{: even and } \left(\frac{x^2 - 2y^2}{p}\right) = 1.$$

Further

$$[\mu] \in C_F(K) \Longleftrightarrow (-1)^{y/2}\left(\frac{ry + x}{p}\right)\left(\frac{2}{x}\right) = 1.$$

We put

$$\begin{cases} E^+ = \{\varepsilon \in \mathfrak{o}^\times \,|\, \varepsilon\text{: totally positive}\}, \\ E^0 = \{\varepsilon \in E^+ \,|\, \varepsilon - 1 \in \mathfrak{f}(K/F)\}, \end{cases}$$

and $e = [E^+ : E^0]$. Then, the right hand side of (3.10) has the following expression for $M = F$:

$$\Theta(\tau;\, K) = e^{-1} \sum_{\substack{\mu = x + 2y\sqrt{2} \\ x \equiv 1 \pmod{4} \\ N_{F/\mathbb{Q}}(\mu) > 0 \\ \mu \bmod E^0}} (\operatorname{sgn} x)(-1)^y \left(\frac{2ry + x}{p}\right)\left(\frac{2}{x}\right) q^{x^2 - 8y^2}. \qquad (3.14)$$

***Case* 3.** $M = E = (= \mathbb{Q}(\sqrt{-2p}))$.

By a calculation similar to that of Case 2, we have the following

$$\Theta(\tau;\, K) = \sum_{\mathfrak{a}} \sum_{4x+1+2y\sqrt{-2p} \in \mathfrak{a}} (-1)^{x+y} q^{\{(4x+1)^2 + 8y^2\}/N_{E/\mathbb{Q}}(\mathfrak{a})}, \qquad (3.15)$$

where $\{\mathfrak{a}\}$ denotes the set of integral ideals of E which are representatives of all square classes in H_E/P_E.

Summing up (3.13), (3.14) and (3.15), we obtain the following theorem which is our main purpose.

Theorem 3.3. *Let p be any prime with $p \equiv 7 \pmod{8}$. Then, the notation and the assumption being kept as above, we have three expression of $\Theta(\tau;\, K)$:*

$$\Theta(\tau;\, K) = \sum_{\mathfrak{a}} \sum_{4x+1+2y\sqrt{-2p} \in \mathfrak{a}} (-1)^{x+y} q^{\{(4x+1)^2 + 8py\}/N_{E/\mathbb{Q}}(\mathfrak{a})} \quad (\textit{via } E)$$

$$= \sum_{\mathfrak{b}} \sum_{4x+1+4y\sqrt{-p}\in\mathfrak{b}^4} (-1)^y q^{\{(4x+1)^2+16py^2\}/N_{k/\mathbb{Q}}(\mathfrak{b})^4} \quad (via\ k)$$

$$= e^{-1} \sum_{\substack{\mu=x+2y\sqrt{2} \\ x\equiv 1 \pmod 4 \\ N_{F/\mathbb{Q}}(\mu)>0 \\ \mu \bmod E^0}} (\operatorname{sgn} x)(-1)^y \left(\frac{2ry+x}{p}\right)\left(\frac{2}{x}\right) q^{x^2-8y^2} \quad (via\ F).$$

Let l be an odd prime number satisfying the conditions $\left(\frac{p}{l}\right) = 1$ and $l \equiv 1 \pmod 8$. Then we have $\left(\frac{\varepsilon_p}{l}\right) = 1$ by (3.9), and we have also the following from the theorem above:

$$l = \{(4a+1)^2 + 8pb^2\}/N_{E/\mathbb{Q}}(\mathfrak{a}),$$
$$l = \{(4\alpha+1)^2 + 16p\beta^2\}/N_{k/\mathbb{Q}}(\mathfrak{b})^4,$$
$$l = x^2 - 8y^2, \quad x \equiv 1 \pmod 4, \quad \left(\frac{x^2-8y^2}{p}\right) = 1;$$
$$a(l) = \pm 2.$$

Moreover, we have the following criteria for ε_p to be a quartic residue modulo l which are conclusion.

$$\left(\frac{\varepsilon_p}{l}\right)_4 = 1 \iff a+b: \text{ even}$$
$$\iff \beta: \text{ even}$$
$$\iff (\operatorname{sgn} x)(-1)^y\left(\frac{2ry+x}{p}\right)\left(\frac{2}{x}\right) = 1 \text{ and } x \equiv 1 \pmod 4$$
$$\iff a(l) = 2.$$

For prime p with $p \equiv 3 \pmod 8$, we shall only state the result as a remark.

Remark 3.5. Let $p \equiv 3 \pmod 8$ and $p \neq 3$. Then, the following may be obtained in a way similar to the proof of the above theorem.

$$\Theta(\tau; K) = \sum_{\substack{x,y\in\mathbb{Z} \\ x\equiv 1 \pmod 4}} (-1)^{(x-1)/4+y}\left(\frac{x-2ry}{p}\right) q^{x^2+8y^2}$$

$$= \sum_{\mathfrak{b}} \Big\{ \sum_{\substack{\nu=(\alpha+\beta\sqrt{-1})/2\in\mathfrak{b}^4 \\ N_{F/\mathbb{Q}}(\nu)\equiv 1 \pmod 8 \\ \alpha\equiv 1 \pmod 4}} (-1)^{(\alpha-1)/4+(N_{k/\mathbb{Q}}(\nu)-1)/8} q^{(\alpha^2+p\beta^2)/4N_{k/\mathbb{Q}}(\mathfrak{b})^4}$$

$$+ \sum_{4x+1+4y\sqrt{-p}\in\mathfrak{b}^4} (-1)^y q^{\{(4x+1)^2+16py^2\}/N_{k/\mathbb{Q}}(\mathfrak{b})^4} \Big\}$$

$$= e^{-1} \sum_{\mathfrak{a}} \sum_{\substack{\mu = 4x+1+2y\sqrt{2p} \in \mathfrak{a} \\ N_{F/\mathbb{Q}}(\mu) > 0 \\ \mu \bmod E^0}} (\operatorname{sgn} x)(-1)^{x+y} q^{\{(4x+1)^2 - 8py^2\}/N_{F/\mathbb{Q}}(\mathfrak{a})}.$$

Chapter 4

Dimension formulas in the case of weight 1

Let Γ be a fuchsian group of the first kind. We shall denote by d_1 the dimension of the linear space of cusp forms of weight 1 on the group Γ. It is not effective to compute the number d_1 by means of the Riemann-Roch theorem. Hejhal said in his book ([37]), it is impossible to calculate d_1 using only the basic algebraic properties of Γ. Because of this reason, it is an interesting problem in its own right to determine the number d_1 by some other method.

In this chapter we give some formula of d_1 by using the Selberg trace formula ([2], [14], [15], [41], [45], [46], [47], [48], [104]), and also discuss $d_1 \bmod 2$.

4.1 The Selberg eigenspace $\mathfrak{M}(k, \lambda)$

Let S denote the complex upper half-plane and we put $G = \mathrm{SL}(2, \mathbb{R})$. Consider direct products

$$\widetilde{S} = S \times T, \quad \widetilde{G} = G \times T,$$

where T denotes the real torus. The operation of $(g, \alpha) \in \widetilde{G}$ on $\widetilde{S}$ is represented as follows:

$$\widetilde{S} \ni (z, \phi) \longrightarrow (g, \alpha)(z, \phi) = \left(\frac{az + b}{cz + d}, \phi + \arg(cz + d) - \alpha \right) \in \widetilde{S},$$

where $g = \begin{pmatrix} a & b \\ c & d \end{pmatrix} \in G$. The space $\widetilde{S}$ is a weakly symmetric Riemann space with the $\widetilde{G}$-invariant metric

$$ds^2 = \frac{dx^2 + dy^2}{y^2} + \left(d\theta - \frac{dx}{2y} \right)^2,$$

and with the isometry μ defined by $\mu(z, \phi) = (-\bar{z}, -\phi)$. The $\widetilde{G}$-invariant measure $d(z, \phi)$ associated to the $\widetilde{G}$-invariant metric is given by

$$d(z, \phi) = d(x, y, \phi) = \frac{dx \wedge dy \wedge d\phi}{y}.$$

The ring of $\widetilde{G}$-invariant differential operators on $\widetilde{S}$ is generated by $\dfrac{\partial}{\partial\phi}$ and

$$\widetilde{\Delta} = y^2\left(\frac{\partial^2}{\partial x^2} + \frac{\partial^2}{\partial y^2}\right) + \frac{5}{4}\frac{\partial^2}{\partial \phi^2} + y\frac{\partial}{\partial\phi}\frac{\partial}{\partial x}.$$

Let Γ be a fuchsian group of the first kind not containing the element $\begin{pmatrix} -1 & 0 \\ 0 & -1 \end{pmatrix}$. By the correspondence

$$G \ni g \longleftrightarrow (g, 0) \in \widetilde{G},$$

we identify the group G with a subgroup $G\times\{0\}$ of $\widetilde{G}$, and so the subgroup Γ identify with a subgroup $\Gamma\times\{0\}$ of $\widetilde{G}$. For an element $(g, \alpha) \in \widetilde{G}$, we define a mapping $T_{(g,\alpha)}$ of $L^2(\widetilde{S})$ into itself by $(T_{(g,\alpha)}f)(z, \phi) = f((g, \alpha)(z, \phi))$. For an element $g \in G$, we put $T_{(g,0)} = T_g$. Then we have

$$(T_g f)(z, \phi) = f\left(\frac{az+b}{cz+d}, \phi + \arg(cz+d)\right),$$

where $g = \begin{pmatrix} a & b \\ c & d \end{pmatrix}$. We denote by $\mathfrak{M}_\Gamma(k, \lambda) = \mathfrak{M}(k, \lambda)$ the set of all functions $f(z, \phi)$ satisfying the following conditions:

(i) $f(z, \phi) \in L^2(\Gamma \backslash \widetilde{S})$,
(ii) $\widetilde{\Delta} f(z, \phi) = \lambda f(z, \phi)$, $(\partial/\partial\phi) f(z, \phi) = -ikf(z, \phi)$.

We call $\mathfrak{M}(k, \lambda)$ the *Selberg eigenspace* of Γ. We denote by $S_1(\Gamma)$ the space of cusp forms of weight 1 for Γ and put

$$d_1 = \dim S_1(\Gamma).$$

Then the following equality holds ([38], [45]):

Theorem 4.1. *The notation and the assumption being as above, we have*

$$\mathfrak{M}\left(1, -\frac{3}{2}\right) = \{e^{-i\phi}y^{1/2}F(z) \mid F(z) \in S_1(\Gamma)\},$$

and hence

$$d_1 = \dim \mathfrak{M}\left(1, -\frac{3}{2}\right). \tag{4.1}$$

Proof. For each $F(z) \in S_1(\Gamma)$ we denote $f(z, \phi)$ on $\widetilde{S}$ by

$$f(z, \phi) = e^{-i\phi} y^{1/2} F(z). \tag{4.2}$$

Then the function $f(z, \phi)$ satisfies the conditions:

1) $f(g(z, \phi)) = f(z, \phi)$ for all $g \in \Gamma$;

2) $(\partial/\partial\phi) f(z, \phi) = -i f(z, \phi)$;

3) $\widetilde{\Delta} f(z, \phi) = -(3/2) f(z, \phi)$ by regularity of $F(z)$ in S;

4) Since $y^{1/2}|F(z)|$ is bounded on S,

$$\begin{aligned} \|f\| &= \frac{1}{2\pi} \int_{\Gamma\backslash\widetilde{S}} \left| e^{-i\phi} y^{1/2} F(z) \right|^2 \frac{dxdyd\phi}{y^2} \\ &= \int_{\Gamma\backslash S} \left| y^{1/2} F(z) \right|^2 \frac{dxdy}{y^2} < \infty. \end{aligned}$$

Therefore, by 1)–4), the function $f(z, \phi)$ belongs to $\mathfrak{M}(1, -(3/2))$. We now prove conversely that any function in $\mathfrak{M}(1, -(3/2))$ must be of the form (4.2) with $F(z) \in S_1(\Gamma)$. Let $f(z, \phi)$ be a function in $\mathfrak{M}(1, -(3/2))$. Put

$$F(z) = e^{i\phi} y^{-1/2} f(z, \phi).$$

Then the Γ-invariance of $f(z, \phi)$ is equivalent to a transformation law for $F(z)$:

$$F(g(z)) = (cz + d) F(z)$$

for all $g = \begin{pmatrix} a & b \\ c & d \end{pmatrix} \in \Gamma$. Therefore, it is sufficient for the proof of the latter half of our theorem, to show that $F(z)$ is holomorphic and vanishes at every cusp of Γ.

Let $\mathfrak{g}$ be the *Lie algebra* of $\mathrm{SL}_2(\mathbb{R})$ $(= G)$. Then we can take the basis $\mathfrak{a}$ of $\mathfrak{g}$ such that the *Lie derivatives* associated with the elements of $\mathfrak{a}$ are given by the following invariant differential operators:

$$\begin{cases} X = y\cos 2\phi \dfrac{\partial}{\partial x} - y \sin 2\phi \dfrac{\partial}{\partial y} + \dfrac{1}{2}(\cos 2\phi - 1)\dfrac{\partial}{\partial \phi}, \\ Y = y \sin 2\phi \dfrac{\partial}{\partial x} - y\cos 2\phi \dfrac{\partial}{\partial y} + \dfrac{1}{2} \sin 2\phi \dfrac{\partial}{\partial \phi}, \\ \Phi = \dfrac{\partial}{\partial \phi}. \end{cases}$$

It is easy to see that

$$\widetilde{\Delta} = \left(X + \frac{1}{2}\Phi\right)^2 + Y^2 + \phi^2.$$

Now we put

$$A^- = 2\left(X + \frac{1}{2}\Phi\right) + 2iY.$$

Then the function $F(z)$ is holomorphic on S if and only if

$$A^- f(z, \phi) = 0. \tag{4.3}$$

To prove (4.3), first note that the operation of A^- depends only on the representations of the Lie algebra $\mathfrak{g}$. Let $L^2_d(\Gamma \setminus G)$ be the discrete part of the space $L^2(\Gamma \setminus G)$. Then $f \in L^2_d(\Gamma \setminus G)$. Let

$$L^2_d(\Gamma \setminus G) = \sum_i V_i$$

be the irreducible splitting of the space $L^2_d(\Gamma \setminus G)$ and put

$$f = \sum_i f_i \quad (f_i \in V_i).$$

Then, if $f_i \neq 0$, we have

$$\widetilde{\Delta} f_i = -\frac{3}{2} f_i, \quad \frac{\partial}{\partial \phi} f_i = -\sqrt{-1} f_i.$$

Therefore, each subspace V_i such that $f_i \neq 0$ is isomorphic to the space H_1 of the irreducible representation of the limit of discrete series. Hence it is sufficient for the proof of (4.3), to show that for any highest weight vector φ in H_1,

$$A^- \varphi = 0. \tag{4.4}$$

For example, by Lemma 5.6 in [55], the relation (4.4) is well known.

Next we shall see the condition for $F(z)$ at every cusp of Γ. Let s be a cusp of Γ. We may assume that $s = \infty$ and the intersection of a fundamental domain for Γ and a neighborhood of ∞ is the following type

$$\{z = x + iy \mid 0 \leqq x \leqq 1,\ y \geqq M\},$$

where M denotes a positive constant. Then, by the condition $f(z, \phi) \in L^2(\Gamma \setminus \widetilde{S})$, we have

$$\int_M^\infty \left\{ \int_0^1 y|F(z)|^2 dx \right\} \frac{dy}{y^2} < \infty.$$

Let

$$F(z) = \sum_{n=-\infty}^{\infty} a_n e^{2\pi i n z}$$

be the Fourier expansion of F at ∞. Then, we have

$$\begin{aligned}\int_0^1 |F(z)|^2 dx &= \int_0^1 \Big(\sum_n a_n e^{2\pi i n z}\Big)\Big(\sum_m \bar{a}_m e^{-2\pi i m \bar{z}}\Big) dx \\ &= \sum_{n,m} a_n \bar{a}_m \int_0^1 e^{2\pi i(n-m)x - 2\pi(n+m)y} dx \\ &= \sum_n |a_n|^2 e^{-4\pi n y}.\end{aligned}$$

Therefore

$$\int_M^\infty y \left(\sum_n |a_n|^2 e^{-4\pi n y}\right) \frac{dy}{y^2} = \sum_n |a_n|^2 \int_M^\infty y^{-1} e^{-4\pi n y} dy.$$

If $n \leqq 0$, then

$$\int_M^\infty y^{-1} e^{-4\pi n y} dy = \infty.$$

So that $a_n = 0$ for all $n \leqq 0$. □

4.2 The compact case

In this section we suppose that the group Γ has a compact fundamental domain in the upper half-plane S. It is well known that every eigenspace $\mathfrak{M}(k, \lambda)$ defined in Section 4.1 is finite dimensional and orthogonal to each other, and also the eigenspaces span together the space $L^2(\Gamma \setminus \widetilde{S})$. We put $\boldsymbol{\lambda} = (k, \lambda)$. For every invariant integral operator with a kernel function $k(z, \phi; z'\phi')$ on $\mathfrak{M}(k, \lambda)$, we have

$$\int_{\widetilde{S}} k(z, \phi;\, z', \phi') f(z', \phi') d(z', \phi') = h(\boldsymbol{\lambda}) f(z, \phi),$$

for $f \in \mathfrak{M}(k, \lambda)$. Note that $h(\boldsymbol{\lambda})$ does not depend on f so long as f is in $\mathfrak{M}(k, \lambda)$. We also know that there is a basis $\{f^{(n)}\}_{n=1}^{\infty}$ of the space $L^2(\Gamma \setminus \widetilde{S})$ such that each $f^{(n)}$ satisfies the condition (ii) in Section 4.1. Then we put $\boldsymbol{\lambda}^{(n)} = (k, \lambda)$ for such a spectra. We now obtain the following *Selberg trace formula* for $L^2(\Gamma \setminus \widetilde{S})$:

$$\sum_{n=1}^{\infty} h(\boldsymbol{\lambda}^{(n)}) = \sum_{M \in \Gamma} \int_{\widetilde{D}} k(z, \phi;\, M(z, \phi)) d(z, \phi), \tag{4.5}$$

where $\widetilde{D}$ denotes a compact fundamental domain of Γ in $\widetilde{S}$ and $k(z,\phi;\,z',\phi')$ is a point-pair invariant kernel of (a)–(b) type in the sense of Selberg such that the series on the left-hand side of (4.5) is absolutely convergent ([86]). Denote by $\Gamma(M)$ the centralizer of M in Γ and put $\widetilde{D}_M = \Gamma(M)\backslash\widetilde{S}$. Then

$$\sum_{M\in\Gamma}\int_{\widetilde{D}} k(z,\,\phi;\,M(z,\,\phi))d(z,\,\phi) = \sum_{l}\int_{\widetilde{D}_{M_l}} k(z,\,\phi;\,M_l(z,\,\phi))d(z,\,\phi), \tag{4.6}$$

where the sum over $\{M_l\}$ is taken over the distinct conjugacy classes of Γ.

We consider an invariant integral operator on the Selberg eigenspace $\mathfrak{M}(k,\,\lambda)$ defined by

$$\omega_\delta(z,\,\phi;\,z',\,\phi') = \left|\frac{(yy')^{1/2}}{(z-\bar{z}')/2i}\right|^{\delta}\frac{(yy')^{1/2}}{(z-\bar{z}')/2i}e^{-i(\phi-\phi')},\quad (\delta>1).$$

It is easy to see that our kernel ω_δ is a point-pair invariant kernel of (a)–(b) type under the condition $\delta>1$ and vanishes on $\mathfrak{M}(k,\,\lambda)$ for all $k\neq 1$. Since $\Gamma\setminus\widetilde{G}$ is compact, the distribution of spectra (k,λ) is discrete and so we put

$$\begin{aligned}&\mu_1 = -\frac{3}{2},\quad \mu_2,\mu_3,\ldots,\\ &d_\beta = \dim\mathfrak{M}(1,\,\mu_\beta),\quad (\beta=1,2,\ldots).\end{aligned}$$

Then the left-hand side of the trace formula (4.5) equals to $\sum_{\beta=1}^{\infty} d_\beta\Lambda_\beta$, where Λ_β denotes the eigenvalue of ω_δ in $\mathfrak{M}(1,\,\mu_\beta)$. For the eigenvalue Λ_β, using the special eigenfunction

$$f(z,\,\phi) = e^{-i\phi}y^{v_\beta},\quad \mu_\beta = v_\beta(v_\beta-1)-\frac{5}{4},$$

for a spectrum $(1,\,\mu_\beta)$ in $L^2(\widetilde{S})$, we obtain

$$\Lambda_\beta = 2^{2+\delta}\pi\frac{\Gamma(1/2)\Gamma((1+\delta)/2)}{\Gamma(\delta)\Gamma(1+(\delta/2))}\Gamma\left(\frac{\delta-1}{2}+v_\beta\right)\Gamma\left(\frac{\delta+1}{2}-v_\beta\right).$$

If we put $v_\beta = 1/2+ir_\beta$, then

$$\Lambda_\beta = 2^{2+\delta}\pi\frac{\Gamma(1/2)\Gamma((1+\delta)/2)}{\Gamma(\delta)\Gamma(1+(\delta/2))}\Gamma\left(\frac{\delta}{2}+ir_\beta\right)\Gamma\left(\frac{\delta}{2}-ir_\beta\right). \tag{4.7}$$

In general, it is known that the series $\sum_{\beta=1}^{\infty} d_\beta\Lambda_\beta$ is absolutely convergent for $\delta>1$. By the Stirling formula, we see that the above series is also absolutely and uniformly convergent for all bounded δ except

$$\delta = \pm(2v_\beta-1).$$

Now we shall calculate the components of trace appearing in the right-hand side of (4.6) ([41]).

1) Unit class $M = \begin{pmatrix} 1 & 0 \\ 0 & 1 \end{pmatrix}$.

   It is clear that $\omega_\delta(z, \phi; M(z, \phi)) = 1$, and hence

$$J(I) = \int_{\widetilde{D}_M} d(z, \phi) = \int_{\widetilde{D}} d(z, \phi) < \infty.$$

2) Hyperbolic conjugacy classes.

   For the primitive hyperbolic element P, we put

$$g^{-1}Pg = \begin{pmatrix} \lambda_0 & 0 \\ 0 & \lambda_0^{-1} \end{pmatrix}, \quad (g \in G), \quad |\lambda_0| > 1$$

and $\Gamma' = g^{-1}\Gamma g$. Then

$$\Gamma'\left(\begin{pmatrix} \lambda_0 & 0 \\ 0 & \lambda_0^{-1} \end{pmatrix}\right) = g^{-1}\Gamma(P)g.$$

The hyperbolic component is calculated as follows:

$$\begin{aligned} J(P^k) &= \int_{\widetilde{D}_P} \omega_\delta(z, \phi; P^k(z, \phi))\, d(z, \phi) \\ &= \int_{g^{-1}\widetilde{D}_P} \omega_\delta(g(z, \phi); P^k g(z, \phi))\, d(z, \phi) \\ &= \int_{g^{-1}\widetilde{D}_P} \omega_\delta(z, \phi; g^{-1}P^k g(z, \phi))\, d(z, \phi) \\ &= (2\pi)(2^{\delta+1}\sqrt{-1})|\lambda_0^k|^{\delta+1}(\operatorname{sgn}\lambda_0)^k \int_{g^{-1}D_P} \frac{y^{\delta-1}}{(z - \lambda_0^{2k}\bar{z})|z - \lambda_0^{2k}\bar{z}|^\delta}\, dxdy, \end{aligned}$$

where $g^{-1}D_P$ is a fundamental domain of $\Gamma'\left(\begin{pmatrix} \lambda_0 & 0 \\ 0 & \lambda_0^{-1} \end{pmatrix}\right)$ in S. Thus,

$$J(P^k) = (2^{3+\delta}\pi)\frac{\Gamma(1/2)\Gamma((\delta+1)/2)}{\Gamma((\delta+2)/2)} \frac{(\operatorname{sgn}\lambda_0)^k \log|\lambda_0|}{|\lambda_0^{-k} - \lambda_0^k||\lambda_0^{-k} + \lambda_0^k|^\delta}.$$

Let $\{P_\alpha\}$ be a complete system of representatives of the primitive hyperbolic conjugacy classes in Γ and let $\lambda_{0,\alpha}$ be the eigenvalue ($|\lambda_{0,\alpha}| > 1$) of representative P_α. Then, the hyperbolic component $J(P)$ is expressed by the following

$$\begin{aligned} J(P) &= \sum_{\alpha=1}^{\infty}\sum_{k=1}^{\infty} J(P_\alpha^k) \\ &= \frac{2^{3+\delta}\pi^{3/2}\Gamma((\delta+1)/2)}{\Gamma((\delta+2)/2)} \sum_{\alpha=1}^{\infty}\sum_{k=1}^{\infty} \frac{(\operatorname{sgn}\lambda_{0,\alpha})^k \log|\lambda_{0,\alpha}|}{|\lambda_{0,\alpha}^k - \lambda_{0,\alpha}^{-k}|} |\lambda_{0,\alpha}^k + \lambda_{0,\alpha}^{-k}|^{-\delta}. \end{aligned}$$

3) Elliptic conjugacy classes.

Let ρ, $\bar{\rho}$ be the fixed points of an elliptic element M ($\rho \in S$) and ζ, $\bar{\zeta}$ be the eigenvalues of M. We denote by Φ be a linear transformation which maps S into a unit disk:

$$w = \Phi(z) = \frac{z-\rho}{z-\bar{\rho}}.$$

Then we have $\Phi M \Phi^{-1} = \begin{pmatrix} \zeta & 0 \\ 0 & \bar{\zeta} \end{pmatrix}$ and

$$\frac{Mz-\rho}{Mz-\bar{\rho}} = \frac{\zeta}{\bar{\zeta}} \frac{z-\rho}{z-\bar{\rho}}.$$

The elliptic component is calculated as follows:

$$\begin{aligned} J(M) &= \int_{\widetilde{D}_M} \omega_\delta(z,\, \phi;\, M(z,\, \phi))\, d(z,\, \phi) \\ &= \frac{2^{\delta+1}\sqrt{-1}}{[\Gamma(M):1]} \int_{\bar{S}} \frac{(yy')^{(\delta+1)/2}}{(z-\bar{z}')|z-\bar{z}'|^\delta} e^{-\sqrt{-1}(\phi-\phi')}\, d(z,\, \phi) \\ &\qquad\qquad ((z',\, \phi') = M(z,\, \phi)) \\ &= \frac{8\pi\bar{\zeta}}{[\Gamma(M):1]} \int_{|w|<1} \frac{(1-w\bar{w})^{\delta-1}}{(1-\bar{\zeta}^2 w\bar{w})|1-\bar{\zeta}^2 w\bar{w}|^\delta}\, dudv \\ &\qquad\qquad (w = u+\sqrt{-1}v) \\ &= \frac{16\pi^2\bar{\zeta}}{[\Gamma(M):1]} \int_0^1 \frac{(1-r^2)^{\delta-1}}{(1-\bar{\zeta}^2 r^2)|1-\bar{\zeta}^2 r^2|^\delta}\, dr. \end{aligned}$$

We put

$$I(\delta) = \int_0^1 \frac{(1-r^2)^{\delta-1} r}{(1-\bar{\zeta}^2 r^2)|1-\bar{\zeta}^2 r^2|^\delta}\, dr.$$

Then under the condition $\delta > 0$, the function $\dfrac{\delta(1-r^2)^{\delta-1} r}{1-\bar{\zeta}^2 r^2}$ is Lebesgue integrable on $[0, 1]$. Hence

$$\begin{aligned} \lim_{\delta\to+0} \delta\, I(\delta) &= \lim_{\delta\to+0} \int_0^1 \frac{\delta(1-r^2)^{\delta-1} r}{1-\bar{\zeta}^2 r^2}\, dr \\ &= \lim_{\delta\to+0} \left\{ \left[-\frac{(1-t)^\delta}{2} \frac{1}{1-\bar{\zeta}^2 t} \right]_0^1 + \int_0^1 (1-t)^\delta \left(\frac{1}{1-\bar{\zeta}^2 t} \right)' \frac{dt}{2} \right\} \end{aligned}$$

$$= \frac{1}{2(1-\bar{\zeta}^2)}.$$

Therefore we obtain

$$\lim_{\delta\to+0} \delta\, J(M) = \frac{8\pi^2}{[\Gamma(M):1]} \frac{\bar{\zeta}}{1-\bar{\zeta}^2}.$$

Since M and M^{-1} are not conjugate and $\bar{\zeta}/(1-\bar{\zeta}^2)$ is pure imaginary, we have

$$\lim_{\delta\to+0} \delta\, J(M) + \lim_{\delta\to+0} \delta\, J(M^{-1}) = 0.$$

We conclude that the contribution from elliptic classes to d_1 vanishes.

Now we put

$$\zeta_1^*(\delta) = \sum_{\alpha=1}^{\infty}\sum_{k=1}^{\infty} \frac{(\operatorname{sgn}\lambda_{0,\alpha})^k \log|\lambda_{0,\alpha}|}{|\lambda_{0,\alpha}^k - \lambda_{0,\alpha}^{-k}|} |\lambda_{0,\alpha}^k + \lambda_{0,\alpha}^{-k}|^{-\delta}. \tag{4.8}$$

Then by the trace formula (4.5), the Dirichlet series (4.8) extends to a meromorphic function on the whole δ-plane and has a simple pole at $\delta = 0$ whose residue will appear in (4.9) below. Finally, multiply the both sides of (4.5) by δ and let δ tends to zero, then the limit is expressed, by the above 1), 2) and 3) as follows:

$$\dim \mathfrak{M}\left(1, -\frac{3}{2}\right) = \frac{1}{2} \operatorname*{Res}_{\delta=0} \zeta_1^*(\delta),$$

namely, by (4.1) we have

$$d_1 = \frac{1}{2} \operatorname*{Res}_{\delta=0} \zeta_1^*(\delta). \tag{4.9}$$

Remark 4.1. Let Γ be a fuchsian group of the first kind which contains the element $-I$, and χ a unitary representation of Γ of degree 1 such that $\chi(-I) = -1$. Let $S_1(\Gamma, \chi)$ be the linear space of cusp forms of weight 1 on the group Γ with character χ, and denote by d_1 the dimension of the linear space $S_1(\Gamma, \chi)$. When the group Γ has a compact fundamental domain in the upper half-plane S, we have the following dimension formula in the same way as in the case $\Gamma \not\ni -I$:

$$d_1 = \frac{1}{2} \sum_{\{M\}} \frac{\chi(M)}{[\Gamma(M):\pm I]} \frac{\bar{\zeta}}{1-\bar{\zeta}^2} + \frac{1}{2} \operatorname*{Res}_{s=0} \zeta_2^*(s), \tag{4.10}$$

where the sum over $\{M\}$ is taken over the distinct elliptic conjugacy classes of $\Gamma/\{\pm I\}$, $\Gamma(M)$ denotes the centralizer of M in Γ, $\bar{\zeta}$ is one of the eigenvalues of M, and $\zeta_2^*(s)$ denotes the Selberg type zeta-function defined by

$$\zeta_2^*(s) = \sum_{\alpha=1}^{\infty}\sum_{k=1}^{\infty} \frac{\chi(P_\alpha)^k \log \lambda_{0,\alpha}}{\lambda_{0,\alpha}^k - \lambda_{0,\alpha}^{-k}} |\lambda_{0,\alpha}^k + \lambda_{0,\alpha}^{-k}|^{-s}. \tag{4.11}$$

Here $\lambda_{0,\alpha}$ denotes the eigenvalue ($\lambda_{0,\alpha} > 1$) of representative P_α of the primitive hyperbolic conjugacy classes $\{P_\alpha\}$ in $\Gamma/\{\pm I\}$.

4.3 The Arf invariant and d_1 mod 2

The purpose of this section is to prove that $d_1 \bmod 2$ is just the Arf invariant of some quadratic form over a field of characteristic 2.

4.3.1 *The Arf invariant of quadratic forms* mod 2

Let V be a vector space of dimension m over a field F of characteristic 2, Q a *quadratic form* on V. Then the associated polar form

$$B(x, y) = Q(x + y) + Q(x) + Q(y)$$

is alternating bilinear form. Let $x_1, \ldots, x_m$ be a symplectic basis of V with respect to B. It is known that the quadratic form $Q(x)$ is equivalent to

$$\sum_{i=1}^{n}\{Q(x_i)a_i^2 + a_i a_{n+i} + Q(x_{n+i})a_{n+i}^2\} + \sum_{i=2n+1}^{m} Q(x_i)a_i^2$$

for $x = \sum_{i=1}^{m} a_i x_i \in V$. By the radical of V we mean the subspace

$$\operatorname{rad} V = \{x \in V \mid B(x, V) = 0\}.$$

We shall say that V is a completely regular space if $\operatorname{rad} V = \{0\}$. We now define the Arf invariant of $Q(x)$ ([3]). Take a 2-dimensional completely regular space U over F and a basis x_1, x_2 for U. Thus

$$U = Fx_1 + Fx_2.$$

Define a multiplication on these basis elements by the following relations:

$$x_1^2 = x_1 \otimes x_1 = Q(x_1),$$
$$x_2^2 = x_2 \otimes x_2 = Q(x_2),$$
$$x_1x_2 + x_2x_1 = B(x_1, x_2) \qquad (= 1).$$

Here we put

$$\omega = x_1x_2, \qquad \theta = x_1.$$

Then we obtain the quaternion algebra $C(U)$ with respect to U:

$$C(U) = F \cdot 1 + F \cdot \theta + F \cdot \omega + F \cdot \theta\omega.$$

It is clear that

$$\theta^2 = a, \quad \omega^2 = \omega + ac, \quad \theta\omega + \omega\theta = \theta, \quad \theta\omega\theta^{-1} = \omega + 1,$$

where $a = Q(x_1)$ $(\neq 0)$ and $c = Q(x_2)$. Therefore, in the separable quadratic field $F(\omega)$ over F, we have the norm

$$N(\alpha + \beta\omega) = \alpha^2 + \alpha\beta + ac\beta^2$$

for every α, β in F. Let F^+ be the additive group of F, and φ a homomorphism

$$\phi : F^+ \ni e \longrightarrow e^2 + e \in F^+,$$

and put

$$\Delta(U) = Q(x_1)Q(x_2) \quad (= ac = N(\omega)).$$

Then we call the class $\Delta(U) \bmod \phi(F^+)$ the *Arf invariant* of U. In general, let

$$V = \mathop{\perp}_{i=1}^{n} U_i \perp \operatorname{rad} V$$

be the orthogonal splitting of the space V into 2-dimensional completely regular subspaces $U_1, \ldots, U_n$. Put

$$\Delta(V) = \sum_{i=1}^{n} \Delta(U_i).$$

Then it is obvious that for a symplectic basis $\{x_1, \ldots, x_m\}$ of V,

$$\Delta(V) = \sum_{i=1}^{n} Q(x_i)Q(x_{n+i}).$$

Now the class $\Delta(V) \bmod \phi(F^+)$ does not depend on the symplectic basis chosen and is called the Arf invariant of Q or the pseudo-discriminant of Q, and is denoted by $\bar{\Delta}(Q)$. In this situation, we have

Theorem 4.2.[1] *Let F be a perfect field, and let V be a completely regular space, so that $m = 2n$. Then the following assertions hold:*

(1) *Two nondegenerate quadratic forms $Q_1(x)$, $Q_2(x)$ on V are equivalent if and only if $\bar{\Delta}(Q_1) = \bar{\Delta}(Q_2)$.*

(2) $Q(x) = \sum_{i=1}^{n} x_i x_{n+i} + \nu(x_n^2 + x_{2n}^2)$; *and therefore* $\bar{\Delta}(Q) = \nu^2$.

[1]For the proof, see Dye ([19]).

4.3.2 *The Atiyah invariant on spin structures*

Let M be a smooth closed oriented surface of genus g and $\mathbb{F}_2$ the 2-element field. We write H_1 and H^1 for $H_1(UM, \mathbb{F}_2)$ and $H^1(UM, \mathbb{F}_2)$ respectively. Let UM be the principal tangential S^1-bundle of M. Then $\widetilde{H}_1$ and $\widetilde{H}^1$ mean $H_1(UM, \mathbb{F}_2)$ and $H^1(UM, \mathbb{F}_2)$ respectively. Then the sequences

$$0 \longrightarrow F_2 \longrightarrow \widetilde{H}_1 \longrightarrow H_1 \longrightarrow 0,$$

$$0 \longrightarrow H^1 \longrightarrow \widetilde{H}^1 \xrightarrow{\delta} F_2 \longrightarrow 0$$

are exact. A spin structure of M is a cohomology class $\xi \in \bar{H}^1$ whose restriction to each fiber is the generator of $\mathbb{F}_2$: $\delta(\xi) = 1$. We denote by Φ the set of spin structures of M. Let a be any homology class in H_1 and let $\widetilde{a}$ be the canonical lifting of a to $\widetilde{H}_1$ (see [57, p. 368]). If a, b are in H_1, then we have

$$\widetilde{(a+b)} = \widetilde{a} + \widetilde{b} + (a \cdot b)z,$$

where z denotes the generator of $\mathbb{F}_2$ as the fiber class and $a \cdot b$ denotes the intersection number of a, b. We define a quadratic form on the symplectic space H_1 over $\mathbb{F}_2$ as a function $\omega : H_1 \to \mathbb{F}_2$ such that

$$\omega(a+b) = \omega(a) + \omega(b) + a \cdot b.$$

Now for $\xi \in \Phi$, we put

$$\omega_\xi(a) = \langle \xi, \widetilde{a} \rangle, \quad a \in H_1,$$

where $\langle \, , \rangle$ denotes the dual pairing of $\widetilde{H}^1$ and $\widetilde{H}_1$. Then the function ω_ξ is a quadratic form on H_1 in the above sense. Indeed, since $\langle \xi, z \rangle = 1$, we have

$$\begin{aligned} \omega_\xi(a+b) &= \langle \xi, \widetilde{a+b} \rangle \\ &= \langle \xi, \widetilde{a} + \widetilde{b} + (a \cdot b)z \rangle \\ &= \langle \xi, \widetilde{a} \rangle + \langle \xi, \widetilde{b} \rangle + (a \cdot b)\langle \xi, z \rangle \\ &= \omega_\xi(a) + \omega_\xi(b) + a \cdot b. \end{aligned}$$

Let Ω be the set of quadratic forms on H_1. Then, D. Johnson proved in [57]:

Lemma. *The mapping $\xi \to \omega_\xi$ gives a bijection from Φ to Ω.*

Next we give the Arf invariant of ω_ξ. For the canonical lifting $\widetilde{a}$ of a in H_1, the mapping on $\widetilde{H}^1$

$$\widetilde{a} : x \longrightarrow \langle x, \widetilde{a} \rangle$$

is linear and we denote by $\bar{a}$ the restriction of $\widetilde{a}$ to $\varPhi$. Let a_i, b_i $(i = 1, \dots, g)$ be a symplectic basis of H_1, i.e.,

$$a_i \cdot a_j = b_i \cdot b_j = 0, \qquad a_i \cdot b_j = \delta_{ij},$$

where δ_{ij} denotes the Kronecker symbol. We put

$$\alpha = \sum_{i=1}^{g} \bar{a}_i \bar{b}_i.$$

Then

$$\begin{aligned} \alpha(\xi) &= \sum_{i=1}^{g} \bar{a}_i(\xi) \bar{b}_i(\xi) = \sum_{i=1}^{g} \langle \xi, \widetilde{a}_i \rangle \langle \xi, \widetilde{b}_i \rangle \\ &= \sum_{i=1}^{g} \omega_\xi(a_i) \omega_\xi(b_i). \end{aligned}$$

Therefore, $\alpha(\xi) \bmod 2$ is the Arf invariant of ω_ξ.

From now on we consider the surface M as a closed Riemann surface of genus g and introduce the Atiyah invariant on M ([5], [71]). Let K be a canonical line bundle on M, and denote by $S(M)$ the set of holomorphic line bundles L on M such that $L \otimes L \cong K$. The elements of $S(M)$ are called theta-characteristic of M. Let D be a divisor on M and let $\mathscr{L}(D)$ denote the space of meromorphic functions f on M such that $D + (f) \geqq 0$. We define the complete linear system of D by

$$|D| = \{ D + (f) \mid f \in \mathscr{L}(D) \}.$$

Then, we have

$$\dim |D| = \dim \mathscr{L}(D) - 1.$$

Let L be the associated line bundle to an effective divisor D and let $\varGamma(L)$ denote the space of holomorphic section of L. Then, since $|D|$ is the projective space associated to $\varGamma(L)$, we have

$$\dim |D| = \dim \varGamma(L) - 1.$$

Theorem 4.3. *The notation being as above, we have the following assertions.*

(1) *For each theta-characteristic L of M,* $\dim \Gamma(L) \bmod 2$ *is stable under deformations of M and L.*
(2) *The set Φ for M corresponds bijectively to the set of isomorphism classes in $S(M)$.*
(3) $\#\{L \in S(M) \mid \dim \Gamma(L) \equiv 0 \bmod 2\} = 2^{g-1}(2^g + 1)$.

The first assertion (1) in Theorem 4.3 is due to Riemann. For the proofs of Theorem 4.3, refer Atiyah ([5]) and Mumford ([71]). By (1) in Theorem 4.3, $\dim \Gamma(L) \bmod 2$ is independent of the choice of the complex structure on M. Now, by combining Lemma and (2) in Theorem 4.3, we have the following diagram:

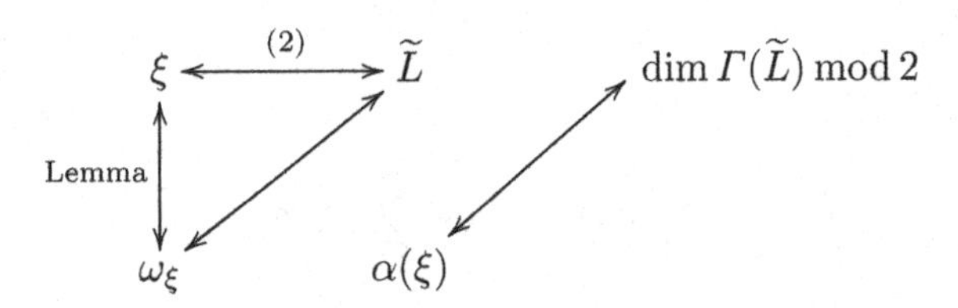

Therefore, intermediating the spin structures $\{\xi\}$ of M, there is a bijection between the isomorphic classes $\{\widetilde{L}\}$ of theta-characteristic and the quadratic forms $\{\omega_\xi\}$ on H_1. It is obvious that the Arf invariant $\alpha(\xi) \bmod 2$ has $2^{g-1}(2^g + 1)$ zeros. Therefore the Arf invariant $\alpha(\xi) \bmod 2$ is equal to the *Atiyah invariant* $\dim \Gamma(\widetilde{L}) \bmod 2$.

4.3.3 *The Arf invariant and* $d_1 \bmod 2$

Let M be a closed Riemann surface of genus g and K a canonical divisor on M. Then, an effective divisor D on M such that $\dim \mathscr{L}(K - D) \neq 0$ is called special. For every special divisor D, we have $0 < \deg D \leqq 2g - 2$. Therefore, the Riemann-Roch theorem says little for special divisors.

Now, let Γ be a fuchsian group of the first kind not containing the element $\begin{pmatrix} -1 & 0 \\ 0 & -1 \end{pmatrix}$, and suppose that the fundamental domain $\Gamma \setminus S$ of Γ is a closed Riemann surface of genus g, where S denotes the upper half-plane. We denote by $P_1, \ldots, P_l$ the point of $\Gamma \setminus S$ corresponding to all the elliptic points of Γ, of order $e_1, \ldots, e_l$, respectively. Let $A_1(\Gamma)$ denote the space of meromorphic automorphic forms of weight 1 with respect to Γ and $S_1(\Gamma)$ the space of holomorphic automorphic forms of weight 1 for Γ. We put

$$d_1 = \dim S_1(\Gamma).$$

For a non-zero element f_0 of $A_1(\Gamma)$, we have

$$\operatorname{div}(f_0) = \frac{1}{2}\operatorname{div}(\omega_{f_0^2}) + \frac{1}{2}\sum_{i=1}^{l}\left(1 - \frac{1}{e_i}\right)P_i \qquad (\omega_{f_0^2} = f_0^2\,dz),$$

and

$$S_1(\Gamma) \cong \mathscr{L}([\operatorname{div}(f_0)]),$$

where $[D] = \sum_i [n_i]P_i$ for $D = \sum_i n_i P_i$ ($n_i \in \mathbb{Q}$, $[n_i]$: Gauss symbol). Put $D_0 = [\operatorname{div}(f_0)]$. Then

$$D_0 = \frac{1}{2}\operatorname{div}(\omega_{f_0^2}) + \sum_{i=1}^{l}\left[\frac{1}{2}\left(1 - \frac{1}{e_i}\right)\right].$$

Therefore we have $\deg D_0 = g - 1$. Hence, under $d_1 \neq 0$, the divisor D_0 is special and

$$\dim \mathscr{L}(D_0) = \dim \mathscr{L}(K - D_0)$$

by the Riemann-Roch theorem. Let L_0 be the associated line bundle of D_0. Then it is obvious that the line bundle L_0 is a theta-characteristic on M. Therefore, we have

$$\begin{aligned} d_1 \bmod 2 &= \dim \mathscr{L}(D_0) \bmod 2 \\ &= \dim \Gamma(L_0) \bmod 2 \\ &= \alpha(\xi_0) \bmod 2 \end{aligned}$$

for the spin structure ξ_0 corresponding to L_0. We have thus the following

Theorem 4.4. *The notation and the assumption being as above, we have the relation*

$$d_1 \bmod 2 = \alpha(\xi_0) \bmod 2.$$

Remark 4.2. We know from Theorem 4.4 that $d_1 \bmod 2$ is the number expressed the topological side of d_1.

Remark 4.3. By Clifford's theorem for special divisors, we have

$$0 \leqq \dim \mathscr{L}(D_0) \leqq \frac{g+1}{2}.$$

But it is impossible to determine $\dim \mathscr{L}(D_0)$ using only the genus g of $\Gamma\backslash S$. For $g = 1$, using the above result and Theorem 4.3 we have $d_1 = 0$.

Now, we may naturally ask the following question:

Can one determine the Arf invariant $\alpha(\xi_0)$ by the basic topological properties of Γ?

4.4 The finite case 1 (: $\Gamma \not\ni -I$)

Let Γ be a fuchsian group of the first kind not containing the element $-I$, and suppose that Γ has a non-compact fundamental domain $\widetilde{D}$ in the space $\widetilde{S}$. Let κ_i be a cusp of Γ and Γ_i be the stabilizer in Γ of κ_i. We take an element $\sigma_i \in \mathrm{SL}(2, \mathbb{R})$ such that $\sigma_i \infty = \kappa_i$ and let γ be a generator of $\Gamma_\infty = \sigma_i^{-1} \Gamma_i \sigma_i$. Then κ_i is called a regular cusp or an irregular cusp according to $\gamma = \begin{pmatrix} 1 & h \\ 0 & 1 \end{pmatrix}$ or $\begin{pmatrix} -1 & h \\ 0 & -1 \end{pmatrix}$ with $h > 0$ respectively. Now, we see that the integral

$$\int_{\widetilde{D}} \sum_{M \in \Gamma} \omega_\delta(z, \phi; M(z, \phi))\, d(z, \phi)$$

is uniformly bounded at a neighborhood of each irregular cusp of Γ, and that by the Riemann-Roch theorem, the number of regular cusps of Γ is even. Therefore we assume for simplicity that $\{\kappa_1, \kappa_2\}$ is a maximal set of cusps of Γ which are regular cusps and not equivalent with respect to Γ. Then the *Eisenstein series* attached to the regular cusp κ_i is defined by

$$E_i(z, \phi; s) = \sum_{\substack{\sigma \in \Gamma_i \backslash \Gamma \\ \sigma_i^{-1}\sigma = \begin{pmatrix} * & * \\ c & d \end{pmatrix}}} \frac{y^s}{|cz + d|^{2s}} e^{-\sqrt{-1}(\phi + \arg(cz+d))} \quad (i = 1, 2), \tag{4.12}$$

where $s = t + \sqrt{-1}r$ with $t > 1$. The series (4.12) has the Fourier expansion at κ_i in the form

$$E_i(\sigma_j(z, \phi); s) = \sum_{m=-\infty}^{\infty} a_{ij,m}(y, \phi; s) e^{2\pi\sqrt{-1}mx}.$$

The constant term $a_{ij,0}(y, \phi; s)$ is given by

$$\begin{aligned} e^{\sqrt{-1}\phi} a_{ij,0}(y, \phi; s) &= a_{ij,0}(y; s) \\ &= \delta_{ij} y^s + \psi_{ij}(s) y^{1-s} \end{aligned}$$

with Kronecker's δ, and

$$\psi_{ij}(s) = -\sqrt{-1}\sqrt{\pi} \frac{\Gamma(s)}{\Gamma(s + (1/2))} \sum_{c \neq 0} \frac{(\operatorname{sgn} c) \cdot N_{ij}(c)}{|c|^{2s}},$$

where $N_{ij}(c) = \# \left\{ 0 \leqq d < |c| \;\middle|\; \begin{pmatrix} * & * \\ c & d \end{pmatrix} \in \sigma_i^{-1} \Gamma \sigma_j \right\}$. We put

$$\Phi(s) = (\, \psi_{ij}(s) \,).$$

Then it is easy to see that the *Eisenstein matrix* $\Phi(s)$ is a skew-symmetric matrix.

Next we define the compact part of the Eisenstein series $E_i(z,\phi;\, s)$ by

$$E_i^Y(z,\phi;\, s) = \begin{cases} E_i(z,\phi;\, s) - a_{ij,0}(\mathrm{Im}\,(\sigma_j^{-1}z),\phi;\, s), & \text{if } \mathrm{Im}\,(\sigma_j^{-1}z) > Y, \\ E_i(z,\phi;\, s), & \text{otherwise}, \end{cases}$$

where Y denotes a sufficiently large number. Then, the following Maass-Selberg relation of our case may be obtained in a way similar to the proof of Theorem 2.3.2 in Kubota [63]: We have the inner product formula

$$\frac{1}{2\pi}\left(E_i^Y(z,\phi;\, s),\ E_j^Y(z,\phi;\, \bar{s}')\right)$$
$$= \frac{Y^{s+s'-1} - \psi_{ij}(s)\overline{\psi_{ij}(\bar{s}')}Y^{-s-s'+1}}{s+s'-1} \qquad (i \neq j). \tag{4.13}$$

We also see that the Eisenstein matrix $\Phi(s)$ converges to a unique unitary matrix $\Phi(s_0)$ when s tends to a point $s_0 = \frac{1}{2} + \sqrt{-1}r_0$. Therefore we have

$$\begin{aligned} \Phi(s_0)\Phi(1-s_0) &= \Phi(s_0)\Phi(\bar{s}_0) \\ &= -\Phi(s_0)\overline{\Phi(s_0)} \\ &= \Phi(s_0)\,{}^t\overline{\Phi(s_0)} = I; \end{aligned}$$

and hence each $E_i(z,\phi;\, s)$ has a meromorphic continuation to the whole s-plane, and the column vector $\mathscr{E}(z,\phi;\, s) = {}^t(E_1,\, E_2)$ satisfies the functional equation

$$\mathscr{E}(z,\phi;\, s) = \Phi(s)\mathscr{E}(z,\phi;\, 1-s).$$

Since Γ is of finite type, the integral operator defined by ω_δ is not completely continuous on $L^2(\Gamma\backslash\widetilde{S})$ in general and the space $L^2(\Gamma\backslash\widetilde{S})$ has the following spectral decomposition

$$L^2(\Gamma\backslash\widetilde{S}) = L_0^2(\Gamma\backslash\widetilde{S}) \oplus L_{\mathrm{sp}}^2(\Gamma\backslash\widetilde{S}) \oplus L_{\mathrm{cont}}^2(\Gamma\backslash\widetilde{S}),$$

where L_0^2 is the space of cusp forms and is discrete, L_{sp}^2 is the discrete part of the orthogonal complement of L_0^2 and L_{cont}^2 is continuous part of the spectra. By using the meromorphic continuation of $E_i(z,\phi;\, s)$, we put

$$\widetilde{H}_\delta(z,\phi\,;\, z',\phi') = \frac{1}{8\pi^2}\sum_{i=1}^{2}\int_{-\infty}^{\infty} h(r)E_i\left(z,\phi\,;\, \frac{1}{2}+\sqrt{-1}r\right)$$
$$\times\, \overline{E_i\left(z',\phi'\,;\, \frac{1}{2}+\sqrt{-1}r\right)}\, dr.$$

Here $h(r)$ denotes the eigenvalue of ω_δ in $\mathfrak{M}(1,\lambda)$ given by (4.4):

$$h(r) = 2^{2+\delta}\pi \frac{\Gamma(1/2)\Gamma((1+\delta)/2)}{\Gamma(\delta)\Gamma(1+(\delta/2))}\Gamma\left(\frac{\delta}{2}+\sqrt{-1}r\right)\Gamma\left(\frac{\delta}{2}-\sqrt{-1}r\right) \tag{4.14}$$

with $\lambda = s(s-1) - \frac{5}{4}$ and $s = \frac{1}{2} + \sqrt{-1}r$. Now we put

$$\kappa_\delta(z,\phi;\, z',\phi') = \sum_{M\in\Gamma} \omega_\delta(z,\phi;\, M(z',\phi'))$$

and

$$\widetilde{\kappa}_\delta = \kappa_\delta - \widetilde{H}_\delta.$$

Then the integral operator $\widetilde{\kappa}_\delta$ is now complete continuous on $L^2(\Gamma\backslash\widetilde{S})$ and has all discrete spectra of κ_δ. Furthermore, an eigenvalue of $f(z,\phi)$ for $\widetilde{\kappa}_\delta$ in $L_0^2(\Gamma\backslash\widetilde{S}) \oplus L_{\mathrm{sp}}^2(\Gamma\backslash\widetilde{S})$ is equal to that for κ_δ and the image of $\widetilde{\kappa}_\delta$ on it is contained in $L^2(\Gamma\backslash\widetilde{S})$. Considering the trace of $\widetilde{\kappa}_\delta$ on $L_0^2(\Gamma\backslash\widetilde{S})$, we now obtain the following modified *trace formula* ([63], [87]):

$$\sum_{n=1}^{\infty} h(\boldsymbol{\lambda}^{(n)}) = \int_{\widetilde{D}} \widetilde{\kappa}_\delta(z,\phi;\, z,\phi)\, d(z,\phi),$$

$$= \int_{\widetilde{D}} \left\{ \sum_{M\in\Gamma} \omega_\delta(z,\,\phi;\, M(z,\,\phi)) - \widetilde{H}_\delta(z,\,\phi;\, z,\,\phi) \right\} d(z,\,\phi)$$

where each of $\boldsymbol{\lambda}^{(n)}$ denotes an eigenvalue corresponding to an orthogonal basis $\{f^{(n)}\}$ of $L_0^2(\Gamma\backslash\widetilde{S})$. We put

$$\int_{\widetilde{D}} \left\{ \sum_{M\in\Gamma} \omega_\delta(z,\,\phi;\, M(z,\,\phi)) - \widetilde{H}_\delta(z,\,\phi;\, z,\,\phi) \right\} d(z,\,\phi)$$

$$= J(I) + J(P) + J(R) + J(\infty),$$

where $J(I)$, $J(P)$, $J(R)$ and $J(\infty)$ denote respectively the identity component, the hyperbolic component, the elliptic component and the parabolic component of the traces. Then the components $J(I)$, $J(P)$ and $J(R)$ are as in Section 4.2 and in the following we shall calculate the component $J(\infty)$ (cf. [46]).

Let $\widetilde{D}_i$ be a fundamental domain of the stabilizer Γ_i of cusp κ_i in Γ. Then we have

$$J(\infty) = \lim_{Y\to\infty} \left\{ \sum_{i=1}^{2} \int_{\widetilde{D}_i^Y} \sum_{\substack{M\in\Gamma_i \\ M\neq I}} \omega_\delta(z,\phi;\, M(z,\phi))\, d(z,\phi) \right.$$

$$- \int_{\widetilde{F}_i^Y} \widetilde{H}_\delta(z,\phi;\, z,\phi)\, d(z,\phi) \Big\},$$

where $\widetilde{D}_i^Y$ denotes the domain consisting of all points (z,ϕ) in $\widetilde{D}_i$ such that $\mathrm{Im}\,(\sigma_i^{-1}z) < Y$ and $\widetilde{F}_i^Y$ denotes the domain consisting of all $(z,\phi) \in \widetilde{D}$ such that $\mathrm{Im}\,(\sigma_i^{-1}z) < Y$ for $i = 1, 2$. For the first half of $J(\infty)$, making use of a summation formula due to Euler-MacLaurin, we have the following

$$\int_{\widetilde{D}_i^Y} \sum_{\substack{M \in \Gamma_i \\ M \neq I}} \omega_\delta\,(z,\phi;\, M(z,\phi))\, d(z,\phi) = 2^2\pi \frac{\Gamma(\frac{1}{2})\Gamma(\frac{\delta+1}{2})}{\Gamma\,(1+\frac{1}{2})} \log Y + \varepsilon(\delta) + o(1)$$

as $Y \to \infty$, where $\varepsilon(\delta)$ denotes a function of δ such that $\lim_{\delta \to +0} \delta\, \varepsilon(\delta) = 0$ (cf. [45]). For the second half of $J(\infty)$, we have

$$\begin{aligned}
&\frac{1}{8\pi^2} \int_{\widetilde{F}_i^Y} h(r) E_i\Big(z,\phi;\, \frac{1}{2} + \sqrt{-1}r\Big) \overline{E_i\Big(z,\phi;\, \frac{1}{2} + \sqrt{-1}r\Big)}\, dr\, d(z,\phi) \\
&= \frac{1}{8\pi^2} \lim_{t \to \frac{1}{2}} \int_{\widetilde{D}} \int_{-\infty}^{\infty} h(r) E_i^Y(z,\phi; t + \sqrt{-1}r)\, \overline{E_i^Y(z,\phi; t + \sqrt{-1}r)}\, dr\, d(z,\phi) \\
&\qquad + o(1) \\
&= \frac{1}{4\pi} \lim_{t \to \frac{1}{2}} \int_{-\infty}^{\infty} h(r) \frac{Y^{2t-1} - \psi_{ij}(s)\overline{\psi_{ij}(s)} Y^{1-2t}}{2t-1} + o(1) \quad \text{(By (4.13))} \\
&= 2^2\pi \frac{\Gamma(\frac{1}{2})\Gamma(\frac{\delta+1}{2})}{\Gamma\,(1+\frac{1}{2})} \log Y \\
&\qquad - \frac{1}{4\pi} \int_{-\infty}^{\infty} h(r) \psi'_{ij}\left(\frac{1}{2} + \sqrt{-1}r\right) \overline{\psi_{ij}\left(\frac{1}{2} + \sqrt{-1}r\right)}\, dr + o(1)
\end{aligned}$$

as $Y \to \infty$ and $t \to \frac{1}{2}$, where $j \neq i$.

Since $\psi_{ij}\left(\frac{1}{2} + \sqrt{-1}r\right) \psi_{ij}\left(\frac{1}{2} - \sqrt{-1}r\right) = 1$, we have

$$\frac{\psi'_{ij}\left(\frac{1}{2} + \sqrt{-1}r\right)}{\psi_{ij}\left(\frac{1}{2} + \sqrt{-1}r\right)} = \frac{\psi'_{ij}\left(\frac{1}{2} - \sqrt{-1}r\right)}{\psi_{ij}\left(\frac{1}{2} - \sqrt{-1}r\right)};$$

hence

$$\begin{aligned}
&\int_{-\infty}^{\infty} h(r) \psi'_{ij}\left(\frac{1}{2} + \sqrt{-1}r\right) \overline{\psi'_{ij}\left(\frac{1}{2} + \sqrt{-1}r\right)}\, dr \\
&= \int_{-\infty}^{\infty} h(r) \frac{\psi'_{ij}}{\psi_{ij}} \left(\frac{1}{2} + \sqrt{-1}r\right) dr.
\end{aligned}$$

By the expression (4.14) of $h(r)$, we obtain

$$h(r) = O\left(\frac{|r|^\delta}{|r| e^{\pi |r|}}\right);$$

and the operator $\widetilde{\kappa}_\delta$ is complete continuous on $L^2(\Gamma\backslash\widetilde{S})$. Therefore we have

$$\lim_{\delta\to+0}\delta\int_{-\infty}^{\infty}h(r)\frac{\psi'_{ij}}{\psi_{ij}}\left(\frac{1}{2}+\sqrt{-1}r\right)dr=0.$$

It is now clear that the above result, with combined with the formula (4.9), proves the following

Theorem 4.5. *Let Γ be a fuchsian group of the first kind not containing the element $-I$ and suppose that the number of regular cusps of Γ is two. Then the dimension d_1 for the space consisting of all cusp forms of weight 1 with respect to Γ is given by*

$$d_1=\frac{1}{2}\operatorname*{Res}_{s=0}\zeta_1^*(s), \tag{4.15}$$

where $\zeta_1^*(s)$ denotes the Selberg type zeta-function defined by (4.8) in Section 4.2.

Remark 4.4. Let Γ be a general discontinuous group of finite type not containing the element $-I$. Then we can prove that in the same way as in the above case, the contribution from parabolic classes to d_1 vanishes.

4.5 The finite case 2 (: $\Gamma\ni -I$)

Let Γ be a fuchsian group of the first kind and assume that Γ contains the element $-I$ and has a non-compact fundamental domain $\widetilde{D}$ in the space $\widetilde{S}$. Let χ be a unitary representation of Γ of degree 1 such that $\chi(-I)=-1$. We denote by $S_1(\Gamma,\chi)$ the linear space of cusp forms of weight 1 on the group Γ with the character χ and by d_1 the dimension of the space $S_1(\Gamma,\chi)$. In this section we shall give similar formula of d_1 when the group Γ is of finite type reduced at infinity and $\chi^2\neq 1$.

Since Γ is of finite type reduced at ∞, ∞ is a cusp of Γ and the stabilizer Γ_∞ of ∞ in Γ is equal to $\pm\Gamma_0$ with $\Gamma_0=\{\left(\begin{smallmatrix}1&m\\0&0\end{smallmatrix}\right):m\in\mathbb{Z}\}$. The Eisenstein series $E_\chi(z,\phi;s)$ attached to ∞ and χ is then defined by

$$E_\chi(z,\phi;s)=\sum_{\substack{M\in\Gamma_\infty\backslash\Gamma\\ M=\left(\begin{smallmatrix}*&*\\c&d\end{smallmatrix}\right)}}\frac{\overline{\chi}(M)y^s}{|cz+d|^{2s}}e^{-\sqrt{-1}(\phi+\arg(cz+d))}, \tag{4.16}$$

where $s=\sigma+\sqrt{-1}r$ with $r>1$. The constant term in the Fourier expansion of (4.16) at ∞ is given by

$$a_0(y,\phi;s)=e^{-\sqrt{-1}\phi}\left(y^s+\psi_\chi(s)y^{1-s}\right),$$

$$\psi_\chi(s) = -\sqrt{-1}\sqrt{\pi}\frac{\Gamma(s)}{\Gamma(s+\frac{1}{2})} \sum_{\substack{c>0 \\ d \bmod c \\ \binom{*\ *}{c\ d}}} \frac{\overline{\chi}(c,d)}{|c|^{2s}}.$$

In the following we only consider the case $\chi\left(\begin{pmatrix} 1 & 1 \\ 0 & 1 \end{pmatrix}\right) = 1$, namely χ is singular. As shown in [45], the parabolic component $J(\infty)$ in the trace formula is given by

$$\begin{aligned} J(\infty) = \lim_{Y\to\infty} \Bigg\{ & \int_0^Y \int_0^1 \int_0^\pi 2 \sum_{\substack{M\in\Gamma \\ M\neq I}} \omega_\delta(z,\phi;\, M(z,\phi))\, d(z,\phi) \\ & - \int_{\widetilde{F}^Y} \widetilde{H}_\delta(z,\phi;\, z,\phi)\, d(z,\phi) \Bigg\} \\ = & -\frac{1}{4\pi}\int_{-\infty}^{\infty} h(r) \frac{\psi'_\chi(\frac{1}{2}+\sqrt{-1}r)}{\psi_\chi(\frac{1}{2}+\sqrt{-1}r)}\, dr - \frac{1}{4} h(0)\psi_\chi(\frac{1}{2}) + \varepsilon(\delta) \end{aligned}$$

as $\lim_{\delta\to+0} \delta\, \varepsilon(\delta) = 0$. When we combine this with the formula (4.10), we are led to the following theorem which is our main purpose in this section.

Theorem 4.6. *Let Γ be a function group of the first kind containing the element $-I$ and suppose that Γ is reduced at infinity. Let χ be a one-dimensional unitary representation of Γ such that $\chi(-I) = -1$, $\chi(\left(\begin{smallmatrix} 1 & 1 \\ 0 & 1 \end{smallmatrix}\right)) =$ 1 and $\chi^2 \neq 1$. We denote by d_1 the dimension of the linear space consisting of cusp forms of weight* 1 *with respect to Γ with χ. Then the dimension d_1 is given by*

$$d_1 = \frac{1}{2}\sum_{\{M\}} \frac{\chi(M)}{[\Gamma(M):+I]} \cdot \frac{\overline{\zeta}}{1-\overline{\zeta}^2} + \frac{1}{2}\operatorname*{Res}_{s=0} \zeta_2^*(s) - \frac{1}{4}\psi_\chi\left(\frac{1}{2}\right), \tag{4.17}$$

where the sum over $\{M\}$ is taken over the distinct elliptic conjugacy classes of $\Gamma/\{\pm I\}$, $\Gamma(M)$ denotes the centralizer of M in Γ, $\overline{\zeta}$ is one of the eigenvalues of M, and $\zeta_2^*(\delta)$ denotes the Selberg type zeta-function defined by (4.11) *in Section 4.2.*

We many call the formulas (4.15) and (4.17) a kind of Riemann-Roch type theorem for automorphic forms of weight 1.

Remark 4.5. For a general discontinuous group Γ of finite type containing the element $-I$, we obtain the contribution from parabolic classes to d_1 in the same way as in the case of reduced at ∞.

4.6 The case of $\Gamma_0(p)$

Let p be a prime number such that $p \equiv 3 \bmod 4$, $p \neq 3$ and let $\Phi_0(p)$ be the group generated by the group $\Gamma_0(p)$ and the element $\kappa = \begin{pmatrix} 0 & -\sqrt{p}^{-1} \\ \sqrt{p} & 0 \end{pmatrix}$, namely, $\Phi_0(p) = \Gamma_0(p) + \kappa\Gamma_0(p)$. Let ε be the Legendre symbol on $\Gamma_0(p)$: $\varepsilon(L) = \left(\frac{d}{p}\right)$ for $L = \begin{pmatrix} a & b \\ c & d \end{pmatrix} \in \Gamma_0(p)$. Since $\varepsilon(\kappa^2) = \varepsilon(-I) = -1$, we can define the odd characters $\varepsilon^{\pm}$ on $\Phi_0(p)$ such that $\varepsilon^{\pm}(\kappa) = \pm\sqrt{-1}$. Then we have

$$S_1(\Gamma_0(p), \varepsilon) = S_1(\Phi_0(p), \varepsilon^+) \oplus S_1(\Phi_0(p), \varepsilon^-).$$

We put

$$\mu_1^{\pm} = \dim S_1(\Phi_0(p), \varepsilon^{\pm}).$$

Then

$$\dim S_1(\Phi_0(p), \varepsilon) = d_1 = \mu_1^+ + \mu_1^-.$$

We denote by $\overline{\Gamma}_0(p)$, $\overline{\Phi}_0(p)$ the inhomogeneous linear transformation group attached to $\Gamma(p)$, $\Phi_0(p)$ respectively. If $\sigma(p)$ is the parabolic class number of $\overline{\Gamma}_0(p)$, then $\sigma(p) = 2$; and if $e_2(p)$, $e_3(p)$ are the number of elliptic classes of order 2, 3 respectively of $\overline{\Gamma}_0(p)$, then

$$e_2(p) = 0, \quad e_3(p) = 1 + \left(\frac{p}{3}\right).$$

Let $\sigma^*(p)$, $e_2^*(p)$, $e_3^*(p)$ denote respectively the number of parabolic classes, the number of elliptic classes of order 2, the number of elliptic classes of order 3 for $\overline{\Phi}_0(p)$. Then we have

$$\sigma^*(p) = \frac{1}{2}\sigma(p) = 1;$$

$$e_3^*(p) = \frac{1}{2}e_3(p) = \frac{1}{2}\left(1 + \left(\frac{p}{3}\right)\right);$$

$$e_2^*(p) = \frac{1}{2}e_2(p) + e_2'(p) = e_2'(p),$$

where $e_2'(p)$ denotes the number of classes of elliptic elements of order 2 of $\kappa\overline{\Gamma}_0(p)$. It is known that

$$e_2'(p) = \left(3 - \left(\frac{2}{p}\right)\right)h = \begin{cases} 4h & \text{if } p \equiv 3 \bmod 8, \\ 2h & \text{if } p \equiv 7 \bmod 8, \end{cases}$$

where h denotes the class number of $\mathbb{Q}(\sqrt{-p})$, which is an odd integer. Let ϑ_2 denote the number of the elements L in $\overline{\Gamma}_0(p)$ such that $\varepsilon^-(\kappa L) = +\sqrt{-1}$. Then, by [75], we have the following

$$\vartheta_2 = \begin{cases} h & \text{if } p \equiv 3 \bmod 8, \\ 0 & \text{if } p \equiv 7 \bmod 8. \end{cases}$$

In the following, we shall calculate the contribution from elliptic elements to $\mu_1^{\pm}$. Let $\{M\}$ be a complete system of representatives of the elliptic conjugacy classes of order 2 in $\overline{\Phi}_0(p)$. Then $\{M\}$ is given by $\left\{\kappa \begin{pmatrix} a & b \\ pb & d \end{pmatrix}\right\}$, where $\left\{\begin{pmatrix} a & b \\ pb & d \end{pmatrix}\right\}$ denotes the representatives of positive definite integral quadratic forms $\begin{pmatrix} a & pb \\ pb & pd \end{pmatrix}$ such that $\det \begin{pmatrix} a & pb \\ pb & pd \end{pmatrix} = p$. Then the result of calculation is given in the following table:

p	$\varepsilon(L)$	The number of elliptic classes of order 2	ζ	$\frac{1}{[\Gamma(M) : \pm I]} \frac{\overline{\zeta}}{1-\overline{\zeta}^2} \varepsilon^{\pm}(\kappa L)$
$p \equiv 3 \bmod 8$	$\varepsilon(L) = 1$	$3h$	$\sqrt{-1}$	$\frac{1}{2} \frac{\sqrt{-1}}{2} (\pm\sqrt{-1}) = \mp\frac{1}{4}$
$p \equiv 3 \bmod 8$	$\varepsilon(L) = -1$	h	$\sqrt{-1}$	$\frac{1}{2} \frac{\sqrt{-1}}{2} (\mp\sqrt{-1}) = \pm\frac{1}{4}$
$p \equiv 7 \bmod 8$	$\varepsilon(L) = 1$	$2h$	$\sqrt{-1}$	$\frac{1}{2} \frac{\sqrt{-1}}{2} (\pm\sqrt{-1}) = \mp\frac{1}{4}$

It is clear that there is no contribution from elliptic classes of order 3 to $\mu_1^{\pm}$. Therefore the contribution from elliptic classes to $\mu_1^{\pm}$ is given by

$$\frac{1}{2} \sum_{\{M\}} \frac{1}{[\Gamma(M) : \pm I]} \frac{\overline{\zeta}}{1-\overline{\zeta}^2} \varepsilon^{\pm}(M) = \mp\frac{1}{4} h.$$

We also have $\psi_\epsilon^{\pm}(1/2) = \mp 1$. Let $\{P_\alpha\}$ be a complete system of representatives of the primitive hyperbolic conjugacy classes in $\overline{\Gamma}_0(p)$ and let $\lambda_{0,\alpha}$ be the eigenvalue ($\lambda_{0,\alpha} > 1$) of representative P_α. We put

$$Z^*(\delta) = \sum_{\alpha=1}^{\infty} \sum_{k=1}^{\infty} \frac{\varepsilon(P_\alpha)^k \log \lambda_{0,\alpha}}{|\lambda_{0,\alpha}^k - \lambda_{0,\alpha}^{-k}|} |\lambda_{0,\alpha}^k + \lambda_{0,\alpha}^{-k}|^{-\delta}.$$

Then, we have consequently the following

$$d_1 = \mu_1^+ + \mu_1^- = \frac{1}{2} \operatorname*{Res}_{\delta=0} Z^*(\delta).$$

Part II

Chapter 5

2-dimensional Galois representations of odd type and non-dihedral cusp forms of weight 1

5.1 Galois representations of odd type

5.1.1 *Artin L-functions and the Artin conjecture*

In this subsection, we shall define anew Artin L-functions for finite Galois extensions and state the famous Artin conjecture.

Let F be a number field and K be a finite Galois extension of F with Galois group $G = \mathrm{Gal}(K/F)$. Let

$$\rho : G \to \mathrm{Aut}_{\mathbb{C}}(V) = \mathrm{GL}(V)$$

be a finite dimensional complex representation of G on an n-dimensional complex vector space V. As always, we build continuity into the definition of 'representation'. For a prime ideal $\mathfrak{p}$ of F and $\mathfrak{P}$ a prime ideal in K above $\mathfrak{p}$, let $D_{\mathfrak{P}}$ denote the *decomposition subgroup* of G corresponding to $\mathfrak{p}$:

$$D_{\mathfrak{P}} = \{\sigma \in G : \sigma(\mathfrak{P}) = \mathfrak{P}\}.$$

The *inertia group* $I_{\mathfrak{P}}$ is the normal subgroup of $D_{\mathfrak{P}}$ consisting of all $\sigma \in D_{\mathfrak{P}}$ such that $\sigma(x) \equiv x \pmod{\mathfrak{P}}$. Let $\sigma_{\mathfrak{P}}$ denote the canonical generator, the Frobenius element at $\mathfrak{P}$, of the cyclic group $D_{\mathfrak{P}}/I_{\mathfrak{P}}$.

The *Artin L-function attached to* ρ is defined to be

$$\begin{aligned} L(s, \rho) &= L(s, \rho, K/F) \\ &= \prod_{\mathfrak{p}} L_{\mathfrak{p}}(s, \rho), \end{aligned}$$

where the product is taken over the nontrivial prime ideals in $\mathfrak{O}_F$, which denotes the ring of integers of F. To define the local factor $L_{\mathfrak{p}}(s, \rho)$, let $V^{I_{\mathfrak{P}}}$ be the subspace of V on which $\rho(I_{\mathfrak{P}})$ acts as the identity. Then the quotient $D_{\mathfrak{P}}/I_{\mathfrak{P}}$ acts on the space $V^{I_{\mathfrak{P}}}$, and we define the Euler factor at $\mathfrak{p}$ to be the polynomial:

$$L_{\mathfrak{p}}(s,\rho) = \det\left(1 - \rho(\sigma_{\mathfrak{P}}) \mid V^{I_{\mathfrak{P}}}(N\mathfrak{p})^{-s}\right)^{-1}$$

for $\mathrm{Re}\, s > 0$. This is well defined and gives the Euler factors at all finite primes.

Each Artin L-function converges absolutely in the half-plane $\mathrm{Re}\, s > 1$ and hence defines an analytic function in that region. From the Brauer induction, it follows that any Artin L-function extends analytically to a meromorphic function on the complex plane $\mathbb{C}$. The Artin L-function $L(s,\rho)$ also satisfies a functional equation of the form

$$L(s,\rho) = \varepsilon(s,\rho)L(1-s,\rho^*),$$

where ρ^* is the contragredient representation to ρ and $\varepsilon(s,\rho)$ is the so-called '*epsilon factor*'. We can now state the famous

Artin conjecture. *If ρ is irreducible and nontrivial, then $L(s,\rho)$ can be analytically continued to an entire function of s.*

This is a very central and important conjecture in number theory. It is part of a general reciprocity law. Nontrivial result in the direction of Artin's conjecture was first obtained by E. Artin. Artin proved his conjecture for monomial representations, those induced from one-dimensional representations of a subgroup. In fact, Artin proved that for such ρ $L(s,\rho)$ is $L(s,\chi)$, a Hecke L-function with character χ. Thus Artin proved a dual form of the fundamental reciprocity law of *abelian class field theory.* Until recently, however, very little was known in general about the entirely of $L(s,\rho)$.

5.1.2 *2-dimensional Galois representations of odd type and the Langlands program*

Let $\bar{\mathbb{Q}}$ denote an algebraic closure of the rational number field $\mathbb{Q}$. Given a finite Galois extension K, there exists a *Galois representation*

$$\rho : \mathrm{Gal}(\bar{\mathbb{Q}}/\mathbb{Q}) \to \mathrm{GL}_n(\mathbb{C}) = \mathrm{GL}(V)$$

with the property that $\mathrm{Gal}(\bar{\mathbb{Q}}/K)$ is the kernel of ρ. Thus we get a faithful representation

$$\rho : \mathrm{Gal}(K/\mathbb{Q}) \to \mathrm{GL}_n(\mathbb{C}).$$

We put $n = 2$. Let c be a complex conjugate, and if the matrix $\rho(c)$ has eigenvalues $+1$, -1, we say that ρ is an *odd representation.* Then we have the following two theorems:

Theorem 5.1 (Weil-Langlands). *Suppose that the representation*

$$\rho : \mathrm{Gal}\,(\bar{\mathbb{Q}}/\mathbb{Q}) \to \mathrm{GL}_2(\mathbb{C}) \tag{5.1}$$

is irreducible and odd with conductor N. Assume that ρ satisfies the following condition

> *(C) The Artin L-function $L(s, \rho \otimes \lambda)$ is an entire function for all twists $\rho \otimes \lambda$ of ρ by one-dimensional representation λ of* $\mathrm{Gal}\,(\bar{\mathbb{Q}}/\mathbb{Q})$.

Then there exists a normalized newform f on $\Gamma_0(N)$ of weight 1 *and character $\varepsilon = \det(\rho)$.*

Theorem 5.2 (Deligne-Serre [17]). *Let f be a normalized newform on $\Gamma_0(N)$ of weight* 1 *and character ε. Then there exists an irreducible odd 2-dimensional Galois representation ρ of* $\mathrm{Gal}\,(\bar{\mathbb{Q}}/\mathbb{Q})$ *with the conductor N and* $\det(\rho) = \varepsilon$ *such that $L(s, f) = L(s, \rho)$.*

In other words, there is a one-to-one correspondence between the set of normalized newforms on $\Gamma_0(N)$ of weight 1 and character ε, and the set of isomorphism classes of irreducible 2-dimensional representations of $\mathrm{Gal}\,(\bar{\mathbb{Q}}/\mathbb{Q})$ with conductor N and determinant odd character ε, under the condition (C). This conjectural correspondence is to be viewed as a generalization of the equivalence, coming from class field theory, of characters of the Galois group of an abelian extension and Hecke characters. Not many different kinds of ρ can occur in (5.1), although those that do arise in a variety of ways. Composing with the projection $\mathrm{GL}_2(\mathbb{C}) \to \mathrm{PGL}_2(\mathbb{C})$, the image of

$$\tilde{\rho} : \mathrm{Gal}\,(\bar{\mathbb{Q}}/\mathbb{Q}) \to \mathrm{PGL}_2(\mathbb{C})$$

must be

$$\tilde{\rho}\left(\mathrm{Gal}\,(\bar{\mathbb{Q}}/\mathbb{Q})\right) = \begin{cases} D_n, & \text{dihedral group,} \\ A_4, & \text{tetrahedral group,} \\ S_4, & \text{octahedral group,} \\ A_5, & \text{icosahedral group.} \end{cases}$$

The above correspondences represent special cases of the *Langlands program.*

Suppose that F is a number field and K is a finite Galois extension of F with Galois group $G = \mathrm{Gal}\,(K/F)$. Let

$$\rho : G \to \mathrm{GL}_n(\mathbb{C})$$

be an n-dimensional representation of G. For each place v of F let ρ_v denote the restriction of ρ to the decomposition group of G at v. The Artin L-function attached to ρ was given by the following

$$L(s,\rho) = \prod_v L(s,\rho_v)$$

extending over all the places of F. If v is unramified in K, and σ_v denotes a Frobenius element over v, then

$$L(s,\rho_v) = \left[\det(I - \rho(\sigma_v)Nv^{-s})\right]^{-1}.$$

For each place v of F let F_v denote the completion of F at v. Let $\mathbb{A}_F$ denote the *adele ring* of F and $G_{\mathbb{A}}$ the *adele group*

$$\mathrm{GL}_n(\mathbb{A}_F) = \prod_v \mathrm{GL}_n(F_v),$$

where the product is a restricted direct one. Let π be any irreducible unitary representation of $G_{\mathbb{A}}$. If π can be realized by right translation operator in the space of automorphic (resp. cuspidal automorphic) forms on GL_n, we call π an *automorphic* (resp. *cuspidal*) representation of GL_n. Then, there is associated to π a family of local representations π_v which is uniquely determined by π and has the following properties:

(1) π_v is irreducible for every v;
(2) π_v is unramified for almost every v;
(3) $\pi = \otimes_v \pi_v$.

Langlands' reciprocity conjecture. *For each Galois representation ρ, there exists an automorphic representation $\pi(\rho)$ of $G_{\mathbb{A}}$ such that $L(s,\rho) = L(s,\pi(\rho))$, where $L(s,\pi(\rho))$ denotes the Hecke-Jacquet-Langlands L-function attached to $\pi(\rho)$. Moreover, if ρ is irreducible and non-trivial, then $\pi(\rho)$ is cuspidal.*

Langlands handled some additional cases when $n = 2$, and later Tunnell was able to deduce an improved result using the methods of Langlands. The following theorem is due to Langlands and Tunnell.

Theorem 5.3. *If ρ is a 2-dimensional complex representation of* $\mathrm{Gal}\,(\bar{F}/F)$ *with solvable image, then the Langlands reciprocity conjecture holds for ρ.*

Example 5.1. $n = 2$ and $F = \mathbb{Q}$. Suppose that $\pi_f = \otimes_p \pi_p$ is generated by the classical modular form

$$f(z) = \sum_{n=1}^{\infty} a_n e^{2\pi i n z}$$

of weight k. The decomposition $\pi_f = \otimes_p \pi_p$ corresponds to the fact that f is an eigenfunction for all Hecke operators T_p. The unramified representation π_p then corresponds to the conjugacy class

$$A_p = \begin{pmatrix} \alpha_p & 0 \\ 0 & \beta_p \end{pmatrix}$$

such that $\det(A_p) = 1$ and $\operatorname{tr}(A_p) = p^{-(k-2)/2} a_p$. In this case, Langlands' reciprocity conjecture (or the Langlands program) can be shown to be equivalent to Artin's conjecture for $L(s, \rho)$ and the hypothetical representation $\pi(\rho)$ corresponds to a cusp form of weight 1. Deligne-Serre proved that all forms of weight 1 can be obtained in this manner (Theorem 5.2).

Now, we think of K as the splitting field of some monic polynomial $h(x)$ with integer coefficients. For almost all primes p, we let σ_p denote the Frobenius element in $\mathrm{Gal}\,(K/\mathbb{Q})$. Recall that the prime splits completely in K if and only if $\sigma = \mathrm{Id.}$, namely, $h(x)$ modulo p splits into distinct linear factors. Let $\mathrm{Spl}(K)$ denote the set of primes p that split completely in K. Then it is clear that

$$\mathrm{Spl}(K) = \{p : \rho(\sigma_p) = I\}.$$

Therefore, under the Langlands reciprocity conjecture, there exists an automorphic representation $\pi = \otimes \pi_p$ of GL_2 such that $A_p = \rho(\sigma_p)$ for almost all p. In particular,

$$\mathrm{Spl}(K) = \{p : A_p(\pi(\rho)) = I\}.$$

Consequently, the Langlands program reduces the problem of determining the set $\mathrm{Spl}(K)$ to the study of automorphic representations of $G_{\mathbb{A}}$.

Example 5.2. $n = 2$ and $F = \mathbb{Q}$. Let $h_p(x)$ be a polynomial reducing the coefficients of $h(x)$ modulo p and $\mathrm{Spl}(h)$ be the set of all primes such that $h_p(x)$ factors into a product of distinct linear polynomials over the finite field $\mathbb{F}_p$. Then the Langlands program brings the following

$$\mathrm{Spl}(h) = \{p : p \nmid D_h, a_p = 2\},$$

where D_h denotes the discriminant of h, $\pi(\rho) = \pi f_\rho$ and $a(p)$ the p-th Fourier coefficient of the cusp form f_ρ of weight 1:

$$f_\rho(z) = \sum_{n=1}^{\infty} a_n e^{2\pi i n z}.$$

5.2 The case of types A_4 and S_4: Base change theory

5.2.1 *Results of Serre-Tate*

Suppose that ρ is an irreducible 2-dimensional *non-dihedral representation* of $\mathrm{Gal}(\bar{\mathbb{Q}}/\mathbb{Q})$ with prime conductor p such that $\varepsilon = \det(\rho)$ is odd.

Theorem 5.4 ([89]). *The notation and the assumptions being as above, we have the following assertions.*

(1) $p \not\equiv 1 \pmod 8$.
(2) If $p \equiv 5 \pmod 8$, *ρ is of type S_4, and the character ε is of order 4 and conductor p.*
(3) If $p \equiv 3 \pmod 4$, *ρ is of type S_4 or A_5, and ε is the Legendre symbol* $\left(\frac{n}{p}\right)$.

Next, we start by taking a Galois extension K of $\mathbb{Q}$ and a prime conductor p. We put $G = \mathrm{Gal}(K/\mathbb{Q})$ and consider the following three cases:

(i) $G \cong S_4$ and $p \equiv 5 \pmod 8$.
(ii) $G \cong S_4$ and $p \equiv 3 \pmod 4$.
(iii) $G \cong A_5$ and $p \equiv 3 \pmod 4$.

An embedding of G in $\mathrm{PGL}_2(\mathbb{C})$ defines a *projective representation* $\tilde{\rho}_K$ of $\mathrm{Gal}(\bar{\mathbb{Q}}/\mathbb{Q})$. Then we have the following

Theorem 5.5 ([89]). *There exists a lifting of $\tilde{\rho}_K$ to $\mathrm{GL}_2(\mathbb{C})$ with prime conductor p and odd determinant if and only if one has the following in the three respective cases above:*

(i) K is the normal closure of a non-real quartic field with discriminant p^3.
(ii) K is the normal closure of a quartic field with discriminant $-p$.
(iii) K is the normal closure of a non-real quintic field with discriminant p^2.

5.2.2 *Base change for* GL_2

Fix E to be a cyclic extension of the number field F, of prime degree ℓ. Then the theory of *base change* describes the correspondence between automorphic representations of the groups $\mathrm{GL}_n(\mathbb{A}_F)$ and $\mathrm{GL}_n(\mathbb{A}_E)$ that reflects the operation of the restriction of Galois representations of W_F to W_E, where W_F (resp. W_E) denotes the *Weil group* of $\bar{F}/F$ (resp. $\bar{E}/E$).

The first results on base change for automorphic forms used the theory of L-functions, and were restricted to the case of quadratic E and GL_2. The introduction of the trace formula to the base change problem is due to H. Saito, who dealt with GL_2 and arbitrary cyclic E using the classical language of automorphic forms, and after that Shintani reformulated Saito's results using group representations. A complete theory of base change for GL_2 and cyclic extensions of degree ℓ was developed by Langlands in a form that was suitable for the later generalization to GL_n $(n > 2)$ by Arthur and Clozel. Hereafter we restrict ourselves to the case $n = 2$.

Suppose that $\pi = \otimes_v \pi_v$ is a cuspidal representation of $\mathrm{GL}_2(\mathbb{A}_F)$, and $\Pi = \otimes_w \Pi_w$ is an automorphic representation of $\mathrm{GL}_2(\mathbb{A}_E)$. Then Π is a base change lift of π, denoted $BC_{E/F}(\pi)$, if for each place v of F and $w \mid v$, the *Langlands parameter* attached to Π_w equals the restriction to W_{E_w} of the Langlands parameter $\sigma_v : W_F \to \mathrm{GL}_2(\mathbb{C})$ of π_v. Then we have the following

Theorem 5.6 (Langlands). *Assume that E/F is a cyclic extension of prime degree. Then*

(1) *(Existence) Every cuspidal representation π of $\mathrm{GL}_2(\mathbb{A}_F)$ has a unique base change lift to $\mathrm{GL}_2(\mathbb{A}_E)$. The lift is itself cuspidal unless E is quadratic over F, and π is monomial or dihedral of the form $\pi(\rho)$ with $\mathrm{Ind}_{W_E}^{W_F} \theta$.*
(2) *(Description of fibers) If two cuspidal representations π and π' have the same base change lift to E, then $\pi' \approx \pi \otimes \omega$ for some character of $F^\times N_{E/F}(\mathbb{A}_E^\times) \setminus \mathbb{A}_F^\times$.*
(3) *(Descent) A cuspidal representation Π of $\mathrm{GL}_2(\mathbb{A}_E)$ equals $BC_{E/F}(\pi)$ for cuspidal π of $\mathrm{GL}_2(\mathbb{A}_F)$ if and only if Π is invariant under the natural action of $\mathrm{Gal}\,(E/F)$.*

Part (1) of this theorem remains true for any extension K/F that can be obtained by successive cyclic extension of prime degree, that is, for any solvable extension.

5.2.3 *The case of types A_4 and S_4*

Langlands proved Artin's conjecture for *tetrahedral* and some *octahedral representations*, and Tunnell extended this to all octahedral representations. These results are based on the Langlands theory of cyclic base change for automorphic representations of GL_2. So the method seems to be restricted to cases where the image of ρ is solvable (cf. Theorem 5.3).

Theorem 5.7 (Langlands). *Assume that ρ is tetrahedral type. Then $\pi(\rho)$ exists.*

Theorem 5.8 (Langlands-Tunnell). *Let ρ be a Galois representation of octahedral type. Then $\pi(\rho)$ exists.*

As the method for proof of the above theorems is similar to each, we shall give the details only for Theorem 5.8

Suppose that for any $\rho : W_F \to \mathrm{GL}_2(\mathbb{C})$ there is a corresponding cuspidal representation $\pi(\rho)$ of $\mathrm{GL}_2(\mathbb{A}_F)$. Then it follows from the definition of base change lifting that $BC_{E/F}(\pi(\rho)) = \pi(\rho_E)$ for any cyclic extension E of F, where $\rho_E = \mathrm{Res}\,\rho|W_E$. This means that if we start with ρ, and want to find candidates for $\pi(\rho)$, then the way to progress is to pick an E such that $\pi(\rho_E)$ is already known to exist, and look among the cuspidal π's such that $BC_{E/F}(\pi) = \pi(\rho_E)$.

In this case, the image of $\rho(W_F)$ in $\mathrm{PGL}_2(\mathbb{C})$ is S_4, and the pull-back of the normal subgroup $A_4 \subset S_4$ is the Weil group W_E of a quadratic extension E of F.

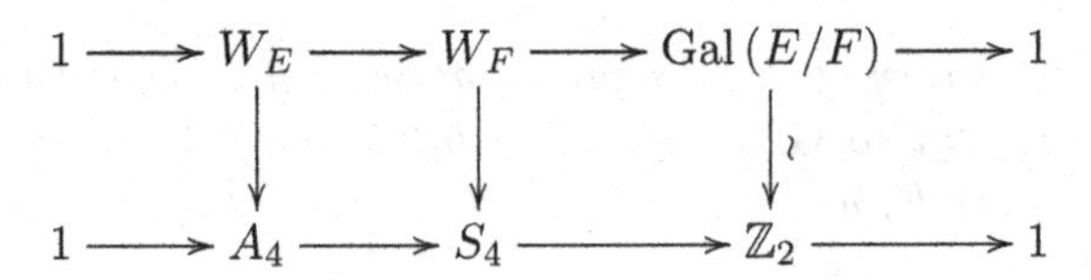

Since ρ_E is now of tetrahedral type, we know that $\pi(\rho_E)$ exists as an irreducible cuspidal representation of $\mathrm{GL}_2(\mathbb{A}_E)$. Since $\pi(\rho_E)$ is invariant under the action of $\mathrm{Gal}\,(E/F)$, $\pi(\rho_E)$ must equal $BC_{E/F}(\pi_i)$ for two irreducible cuspidal representations π_i of $\mathrm{GL}_2(\mathbb{A}_F)$.

Let L/F be a nonnormal cubic subextension of K/F fixed by a 2-Sylow subgroup (of order 8) of S_4. Then if M is the composition in K of L and E, we have the following diagram:

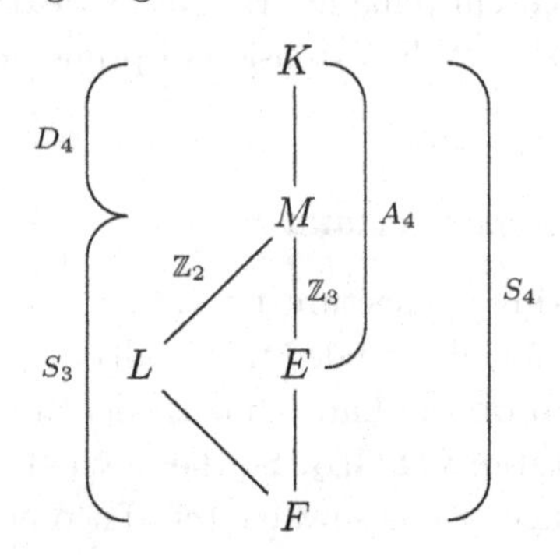

Tunnell's contribution to Theorem 5.8 was to get the following result: There is a unique i such that $BC_{L/F}(\pi_i) = \pi(\rho_L)$. □

5.3 The case of type A_5

The icosahedral case has until now largely been attached using computational methods, where one can hope to construct an explicit weight 1 exotic modular form to deal with any particular case. There is a growing literature on the computational side of the subject, beginning with Buhler [8] and continued by Frey ([23]) and others.

5.3.1 *The first example due to Buhler*

Let us consider the polynomial

$$F(x) = x^5 + 10x^3 - 10x^2 + 35x - 18.$$

This has discriminant $2^6 5^8 11^2$, and the aim is to find the pattern of how $F(x)$ reduces modulo p for $p \neq 2, 5, 11$. The Langlands theory gives conjectures that describe the pattern. Let K be the splitting field of $F(x)$ over $\mathbb{Q}$ and let G be the Galois group. Then we know that K has the conductor 800, $\mathrm{Gal}\,(K/\mathbb{Q}) \cong A_5$ and the ramified primes are 2 and 5 with decomposition groups isomorphic to A_4 and $\mathbb{Z}/5\mathbb{Z}$ respectively.

Theorem 5.9 (Buhler). *There is an icosahedral representation*

$$\rho : \mathrm{Gal}\,(\bar{\mathbb{Q}}/\mathbb{Q}) \to \mathrm{GL}_2(\mathbb{C})$$

of conductor 800 *such that* $L(s, \rho)$ *is an entire function of* s.

In the following we shall outline the proof that was provided by Buhler.

Firstly the quintic polynomial $F(x)$ gives rise to the *icosahedral representation* ρ of conductor 800. The next stage in the proof is the calculation of some of the coefficients of the L-series $L(s, \rho) = \sum a_n n^{-s}$ of a representation ρ attached to $F(x)$. Namely, let $f(z) = \sum a_n q^n$ be the Mellin transform of the L-series $L(s, \rho)$, then the aim is to show that this $f(z)$ coincides with the q-expansion of cusp form of weight 1 and level 800 up to a required number of terms. This is shown as follows.

Let V be the vector space of modular forms of type $(1, \varepsilon, 800)$, i.e., of weight 1, character ε and level dividing 800. The choice of ε is such that in V, there are two non-cuspidal eigenforms of level 100, denoted by g_1, g_2, and there is only one dihedral form of level 100, g_3 which is a cusp form. Each of these forms can 'pushed up' to level 800 by the Atkin-Lehner operator B_d, $d = 1, 2, 4, 8$: $(g|B_d)(z) = g(dz)$.

Let $g_{i,d} = g_i|B_d$. If g is a modular form of type $(1, \varepsilon, N)$, let $\bar{g}$ denote the 'complex conjugate' of g; $\bar{g}$ is of type $(1, \bar{\varepsilon}, N)$ and the Fourier coefficients of $\bar{g}$ are the complex conjugate of the Fourier coefficients of g. For each $i = 1, 2, 3$ and $d = 1, 2, 4, 8$ there is a modular form $h_{i,d}$ of weight 2, level 800 and trivial character such that

$$f\bar{g}_{i,d} \equiv h_{i,d} \pmod{q^{360}},$$

where mod q^M means that the first M terms of the two power series agree. From this and some more result, we can define a function

$$f' = \frac{h_{i,d}}{\bar{g}_{i,d}}$$

that is independent of the choice of i and d. We know that f' is a cusp form of type $(1, \varepsilon, 800)$. It is easy to enumerate the dihedral form of this level and character, and their eigenvalues for the Hecke operator T_3 are unequal to the eigenvalue of f' under the action of T_3. Hence f' is not of the dihedral type. It is not of the tetrahedral or octahedral type follows from that there are no cyclic extension of $\mathbb{Q}$ of degree 3 unramified outside 2 and 5 and hence there are no A_4 extension of $\mathbb{Q}$ unramified outside 2 and 5, and there are exactly three S_4 extension unramified outside 2 and 5 and the corresponding representations have conductors not dividing 800. Therefore there must be an icosahedral form in V.

The proof shows only that the first 360 coefficients of the q-expansion of f' agree with the corresponding initial segment of the power series f obtained from $L(s, \rho)$. If f' is an eigenform for the Hecke operator T_{11}, then f' is the same as f. The proof uses the technique of Odlyzko, Serre and Poitou for bounding discriminants.

Summing up, there exists an icosahedral form of level 800, and Theorem 5.9 follows immediately from the theorem by the result of Deligne-Serre ([17]) together with this result. □

5.3.2 *Icosahedral Artin representations*

The contribution to the problem by Taylor and others is treat infinitely many icosahedral cases using a theoretical approach. More precisely, they proved the following

Theorem 5.10. *Suppose that $\rho : \mathrm{Gal}(\bar{\mathbb{Q}}/\mathbb{Q}) \to \mathrm{GL}_2(\mathbb{C})$ is a continuous irreducible representation and ρ is odd. If ρ is icosahedral, suppose that the*

projective representation $\tilde{\rho}$ of ρ is unramified at 2, the image of a Frobenius element at 2 under $\tilde{\rho}$ has order 3 and $\tilde{\rho}$ is unramified at 5. Then there is a weight 1 newform f such that for all prime p, the p-th Fourier coefficient of f equals the trace of Frobenius at p on the space of coinvariants for the inertia group at p in the representation ρ. In particular, the Artin L-series $L(s,\rho)$ for ρ is Mellin transform of a weight 1 newform and is an entire function of s.

The proof follows a strategy outlined to Wiles by Taylor in 1992 ([10]).

5.4 The Serre conjecture

Let $G_{\mathbb{Q}} = \mathrm{Gal}\,(\bar{\mathbb{Q}}/\mathbb{Q})$ be the absolute Galois group of $\mathbb{Q}$ and let

$$\bar{\rho} : G_{\mathbb{Q}} \longrightarrow \mathrm{GL}_2(\mathbb{F})$$

be a continuous, absolutely irreducible, 2-dimensional with $\mathbb{F}$ a finite field of characteristic p. We say that such a representation is of S-*type*. Serre has conjectured in [90] that such a $\bar{\rho}$ is a modular, i.e., arises from (with respect to the fixed embedding $\iota_p : \bar{\mathbb{Q}} \hookrightarrow \bar{\mathbb{Q}}_p$) a newform f of weight $k(\bar{\rho})$ and level $N(\bar{\rho})$, where $N(\bar{\rho})$ is the Artin conductor of $\bar{\rho}$ (prime to p), and $k(\bar{\rho})$ the Serre weight of $\bar{\rho}$ as defined in [90]. This conjecture was proved by Khare-Wintenberger (Serre's modularity conjecture (I), (II), Invent. Math., **78**, 2009).

For a finite set of primes S of $\mathbb{Q}$ we denote by $G_{\mathbb{Q},S}$ the Galois group of the maximal subfield of $\bar{\mathbb{Q}}$ unramified outside of S. Then we have the following

Theorem (Kisin [59]). *Suppose that*

$$\bar{\rho} : G_{\mathbb{Q},S} \longrightarrow \mathrm{GL}_2(\mathbb{F})$$

is odd and absolutely irreducible. Then $\bar{\rho}$ is modular.

Corollary. *Let*

$$\rho : G_{\mathbb{Q},S} \longrightarrow \mathrm{GL}_2(\mathbb{C})$$

be a continuous, irreducible, odd representation. Then ρ arises from a weight 1 cusp form on $\Gamma_1(N)$ for some $N \geqq 1$. In particular, the Artin L-function $L(\rho,\, s)$ is entire.

Finally the following statement holds: 'Let $\rho : G_{\mathbb{Q}} \longrightarrow \mathrm{GL}_2(\mathbb{C})$ be an odd 2-dimensional complex representation with projective image A_5 and let N be its conductor. Then there exists an eigenform f of weight 1 for $\Gamma_1(N)$ such that ρ is isomorphic to the Galois representation attached to f by Deligne-Serre.' (C. Khare and J.-P. Wintenberger)

5.5 The Stark conjecture in the case of weight 1

5.5.1 *The Stark conjecture*

Let $\mathfrak{a} \neq (1)$ be an integral ideal in $k = \mathbb{Q}(\sqrt{d})$ where $d(< 0)$ is the discriminant of k. If χ is a ray class character of k mod $\mathfrak{a}$, then we may write

$$L(s, \chi) = \sum_{C} \chi(C) Z(s, C),$$

where C runs through the ray classes mod $\mathfrak{a}$ and

$$Z(s, C) = \sum_{\mathfrak{b} \in C} N(\mathfrak{b})^{-s}.$$

Define $g_\chi(z)$ by the Mellin transform, such that

$$(2\pi)^{-s} \Gamma(s) L(s, \chi) = \int_0^\infty y^{s-1} g_\chi(iy) dy, \quad z = x + iy.$$

Then, $g_\chi(z)$ is a modular forms of weight 1 on $\Gamma_1(N)$ with $N = |d| N(\mathfrak{a})$ and we have that

$$L'(0, \chi) = \int_0^\infty g_\chi(iy) \frac{dy}{y}.$$

Now we are led to the following Stark conjecture ([98]).

Conjecture. *Let $f(z)$ be a cusp form of weight* 1 *on $\Gamma_1(N)$. Then*

$$\int_0^\infty f(iy) \frac{dy}{y} = \sum_{j=0}^{n} \rho_j \log \varepsilon_j,$$

where the ε_j are algebraic integers and the ρ_j lie in the field generated over $\mathbb{Q}$ by adjoining the Fourier coefficients of $f(z)$ at ∞.

As an example, let χ be either one of the two cubic ideal class characters of $\mathbb{Q}(\sqrt{-23})$ so that

$$g_\chi(z) = \eta(z)\eta(23z),$$

where $\eta(z)$ denotes the Dedekind eta function. Then we have

$$L'(0, \chi) = \int_0^\infty g_\chi(iy) \frac{dy}{y} = \log \varepsilon_0,$$

where ε_0 is the real root of $x^3 - x - 1 = 0$.

According to the Deligne-Serre theorem, there is a normal extension K of $\mathbb{Q}$ and an irreducible 2-dimensional Galois representation σ of $\mathrm{Gal}\,(K/\mathbb{Q})$

such that the Dirichlet series corresponding to $f(z)$ gives the Artin L-function $L(s, \sigma, K/\mathbb{Q})$. However from the Deligne-Serre theorem, we can expect nothing to solve the problem of explicitly determining the field K by $f(z)$. The conjecture was proved by Stark when K is an abelian extension of k and it aids materially in explicitly determining K from $f(z)$.

In [12], Chinburg formulated Stark conjecture "over $\mathbb{Z}$" as follows. Let $d = \sum_\sigma d_\sigma \cdot \sigma$ be a finite linear combination of ρ of dimension n and we assume that $\sum_\sigma d_\sigma \cdot \sigma = \sum_\sigma d_\sigma^\rho \cdot \sigma^\rho$ for any $\rho \in \mathrm{Aut}(\mathbb{C}/\mathbb{Q})$. We define $L'(s, d) = \sum_\sigma d_\sigma L'(s, \sigma)$ and $L'_d(s) = \sum_\sigma d_\sigma \cdot L'(s, \sigma) pr_\sigma$ where $pr_\sigma = \sum_{g \in \mathrm{Gal}\,(K/\mathbb{Q})} L'_d(0) v_0 = \sum_{v \in S_\infty} \log \|e(d)\|_v \cdot v$ where S_∞ is the set of infinite place of K, $\| \, \|_v$ is the normalized absolute value for $v \in S_\infty$ and v_0 is a fixed embedding of K into $\mathbb{C}$.

Moreover, Tanigawa gave an example for 2-dimensional representation of S_4-type ([103]). He considered the space of cusp forms of weight 1 on $\Gamma_0(283)$ with the character $\left(\dfrac{-283}{*}\right)$. This space has one primitive form h of S_3-type and two primitive forms f and f^τ of S_4-type, where τ is a complex conjugate. Now let V and W be Galois representations attached to f and h respectively. Then $L'_d(0)$ is generated by a linear combination of $L'_d(0)$ for the following d:

(i) $d = \delta V + \delta^\tau V^\tau$ for $\delta \in D^{-1}_{\mathbb{Q}(\sqrt{-2})}$,
(ii) $d = W$,
(iii) $d = \dfrac{1}{4}(V + V^\tau) + \dfrac{1}{2} W$,

here D_k is the different of the field k. Furthermore, he gave the minimal polynomial of $e(d)$ for the above d and checked that $e(d)$ is indeed a real unit in K.

5.5.2 *The value of* $L\left(\frac{1}{2}, \varepsilon\right)$

Let ε be an abelian character of a class group of a complex quadratic extension of a totally real field and $L(s, \varepsilon)$ the Artin L-function associated with ε. Then Moreno asked the values of $L\left(\dfrac{1}{2}, \varepsilon\right)$ and obtained the following result ([69]).

Let σ be an irreducible 2-dimensional linear representations of $G =$

Gal $(\bar{\mathbb{Q}}/\mathbb{Q})$ and $L(s, \sigma)$ be the Artin L-function associated with σ. We put

$$L(s, \sigma) = (2\pi)^{-1}\Gamma(s)\sum_{n=1}^{\infty} a(n)n^{-s}.$$

If σ is a lifting of the projective representation $\tilde{\sigma}$ of G and $\mathrm{Im}(\tilde{\sigma}) = S_3$, then by the theorem of Hecke, the function

$$f(z) = \sum_{n=1}^{\infty} a(n)e^{2\pi inz}$$

is normalized newform on $\Gamma_0(N)$ of weight 1 and character $\varepsilon(= \det(\sigma))$, where N denotes the conductor of σ.

On the other hand, let $E(s, z, \Gamma_0(N))$ be the non-holomorphic Eisenstein series for $\Gamma_0(N)$ corresponding to the cusp at ∞. The Maclaurin expansion of $E(s, z, \Gamma_0(N))$ about $s = 0$ is

$$E(s, z, \Gamma_0(N)) = f^*(z)s + O(s^2),$$

where $f^*(z)$ is a real analytic automorphic form for $\Gamma_0(N)$ with the eigenvalue 1/4 for the Laplacian $-y^2\left(\frac{\partial^2}{\partial x^2} + \frac{\partial^2}{\partial y^2}\right)$. Then he obtained that

$$\Lambda_k\left(\frac{1}{2}\right) L\left(\frac{1}{2}, \sigma\right) c\left(\frac{1}{2}\right) = \langle f^* \cdot f, f\rangle,$$

where $\langle\, , f\rangle$ denotes the Petersson inner product, k the complex quadratic field corresponding to ε, $\Lambda_k(s)$ the *Dedekind zeta-function* of k and

$$c\left(\frac{1}{2}\right) = \prod_{\substack{p\mid N \\ p:\ \text{prime}}} \frac{(1-\varepsilon(p)p^{-1/2})(1-a(p)p^{-1/2})}{(1+p^{-1/2})(1-a(p)^2p^{-1/2})}.$$

Now we ask the following non-dihedral problem. We suppose that $\mathrm{Im}(\tilde{\sigma}) = S_4$. Then, by the theorem of Weil-Langlands-Tunnell, the function $f(z)$ corresponding to $L(s, \sigma)$ by the Mellin transformation is a normalized newform on $\Gamma_0(N)$ of weight 1 and character ε. We may naturally ask the following question:

Can one express the value of $L\left(\frac{1}{2}, \sigma\right)$ as a sum of values of a non-holomorphic modular form at special points?

Chapter 6

Maass cusp forms of eigenvalue 1/4

6.1 Maass cusp forms and Galois representations of even type

6.1.1 *Maass forms of weight zero*

If we consider modular forms without the holomorphy condition but insist that our function is an eigenfunction of the *non-Euclidean Laplacian*

$$\Delta = -y^2 \left(\frac{\partial^2}{\partial x^2} + \frac{\partial^2}{\partial y^2} \right),$$

we arrive at the notion of a real analytic form.

A *Maass form of weight zero* for $\Gamma_0(N)$ is a complex-valued function $f(z)$ on the upper half-plane S satisfying the conditions:

1) f is smooth;
2) f is an eigenfunction of the Laplacian Δ;
3) $f\left(\frac{az+b}{cz+d}\right) = f(z)$ for all $\begin{pmatrix} a & b \\ c & d \end{pmatrix} \in \Gamma_0(N)$;
4) f has at most polynomial growth at each cusp of $\Gamma_0(N)$.

We let $M_\Delta(\lambda, \Gamma_0(N))$ denote the space of Maass forms for $\Gamma_0(N)$ with eigenvalue λ. Let $f \in M_\Delta(\lambda, \Gamma_0(N))$. Then we have

$$f(z) = f(x, y) = \sum_n a_n(y) e^{2\pi i n x},$$

since $f(z+1) = f(z)$. Moreover, by the differential equation which f satisfies and our given growth condition, we have a Fourier expansion of the form

$$f(x, y) = a_0 y^s + a_0' y^{1-s} + \sum_{n \neq 0} a_n \sqrt{y} K_{ir}(2\pi |n| y) e^{2\pi i n x},$$

where

$$K_{ir}(2\pi|n|y) = \frac{1}{2}\int_{-\infty}^{\infty} e^{-y\cosh t - irt}dt$$

with $\lambda = \frac{1}{4} + r^2$.

Also, either $a_n = a_{-n}$ or $a_n = -a_{-n}$, according to which we say that f is even or odd respectively. We say that f is a Maass cusp form if $a_0 = a_0' = 0$ and similarly for every other cusp, and denote by $S_\Delta(\lambda, \Gamma_0(N))$ the subspace of cusp forms in $M_\Delta(\lambda, \Gamma_0(N))$.

The *Ramanujan-Petersson conjecture* for Maass forms then proposes that $a_n = O(n^\varepsilon)$ for any $\varepsilon > 0$, and *Selberg's eigenvalue conjecture* is that $\lambda \geqq 1/4$, or equivalently r is real and not purely imaginary. Langlands interprets the Selberg conjecture as a Ramanujan-Petersson conjecture 'at infinity'.

For every Maass form f, we associate a Dirichlet series

$$L(s, f) = \sum_{n=1}^{\infty} a_n n^{-s}.$$

Maass proved that the series $L(s, f)$ extends to a meromorphic function for all $s \in \mathbb{C}$ analytic everywhere except possibly at $s = 0$ and $s = 1$, and satisfies a functional equation.

6.1.2 *Maass forms with weight*

Let us fix a discrete subgroup Γ of $\mathrm{SL}_2(\mathbb{R})$. We consider functions on the extended upper half-plane S^* which satisfy the following

1) $f\left(\frac{az+b}{cz+d}\right) = \left(\frac{c\bar{z}+d}{|cz+d|}\right)^k f(z)$ for all $\begin{pmatrix} a & b \\ c & d \end{pmatrix} \in \Gamma$;

2) f is an eigenfunction of

$$\Delta_k = -y^2\left(\frac{\partial^2}{\partial x^2} + \frac{\partial^2}{\partial y^2}\right) + iky\frac{\partial}{\partial x};$$

3) a growth condition of the form

$$f(x+iy) = O\left(y^\ell\right)$$

for some $\ell > 0$ as y tends to infinity.

We call such f a *Maass form of weight* k. If g is a classical modular form of weight k, then $y^{\frac{k}{2}}g(z)$ is a Maass form of weight k with eigenvalue $\frac{1}{4}k(2-k)$. Therefore the study of Maass forms includes the study of modular forms from this perspective.

The set of Maass forms of a fixed weight and eigenvalue is a vector space over $\mathbb{C}$, and we can define an involution acting on this space given by

$$\iota : f(z) \mapsto f(-\bar{z}).$$

A form is called *even* if $\iota(f) = f$ and *odd* if $\iota(f) = -f$. Therefore the space of Maass forms decomposes as a direct sum of two subspaces consisting of even forms and odd forms respectively.

The L-series attached to f

$$L(s, f) = \sum_{n=1}^{\infty} a_n n^{-s}$$

extends to an entire function and satisfies a functional equation.

6.1.3 *Galois representations of even type*

Suppose that $\rho : \mathrm{Gal}\,(\bar{\mathbb{Q}}/\mathbb{Q}) \to \mathrm{GL}_2(\mathbb{C})$ is a continuous irreducible representation and that ρ is *even*, that is

$$\rho(\tau) \sim \begin{pmatrix} (-1)^m & 0 \\ 0 & (-1)^m \end{pmatrix} \quad (m = 0, 1)$$

with complex conjugation τ in $\mathrm{Gal}\,(\bar{\mathbb{Q}}/\mathbb{Q})$. Let N be the Artin conductor of ρ and

$$L(s, \rho) = \sum a_n n^{-s}$$

its Artin L-series. Define a function f_ρ on S by the following

$$f_\rho(x + iy) = \sum_{n \neq 0} \frac{(\mathrm{sgn}\, n)^m a_{|n|}}{\sqrt{|n|}} \sqrt{y} K_0(2\pi|n|y) e^{2\pi i n x}.$$

Then we can state the following conjecture.

Casselman conjecture. *The function f_ρ lies in $S_\Delta^m(\frac{1}{4}, \Gamma_1(N))$, where S_Δ^0 (resp. S_Δ^1) denotes the space of even Maass cusp forms (resp. odd Maass cusp forms).*

This is true if and only if the Artin conjecture holds for all the representations $\rho \otimes \chi$, where χ ranges over all Dirichlet characters. One might hope that all newforms in $S_\Delta^m(\frac{1}{4}, \Gamma_1(N))$ can be obtained in this manner (Casselman [11]). This would be the result analogous to that of Deligne-Serre ([17]) for these forms. However the techniques applied by Deligne-Serre seem unlikely to work here.

6.2 Automorphic hyperfunctions of weight 1

6.2.1 *Limits of discrete series*

V. Bargmann classified the irreducible unitary representation of $\mathrm{SL}_2(\mathbb{R})$. In this classification, there are two special cases, called the '*limits of discrete series*', and denoted by D_1^+ and D_1^-. The group action with D_1^+ is

$$D_1^+\begin{pmatrix} a & b \\ c & d \end{pmatrix} f(z) = (-bz+d)^{-1} f\left(\frac{az-c}{-bz+d}\right)$$

for $\begin{pmatrix} a & b \\ c & d \end{pmatrix} \in \mathrm{SL}_2(\mathbb{R})$. The representation D_1^- is obtained by complex conjugation. Also the norm is given by

$$\|f\|^2 = \sup_{y>0} \int_{-\infty}^{\infty} |f(x+iy)|^2 dx.$$

Both of the representations D_1^+ and D_1^- are irreducible, and we have the relation

$$P^{-,0} = D_1^+ \oplus D_1^-,$$

where $P^{-,0}$ denotes one of the *principal series* of $\mathrm{SL}_2(\mathbb{R})$. In addition the representations D_1^+ and D_1^- are not square integrable. D_1^+ is realized in holomorphic weight 1 modular forms of the upper half-plane S.

6.2.2 *Automorphic hyperfunctions of weight* 1

6.2.2.1 *Hyperfunctions of one variable*

In this subsection we briefly summarize the results of hyperfunctions of one variable following [84].

For any open set $D \subset \mathbb{C}$, $H(D)$ denotes the ring of all holomorphic functions on D. Let S be any locally closed subset of $\mathbb{R}$. We denote by $N(S)$ the family of all complex neighborhoods of S, i.e., all open sets of $\mathbb{C}$ containing S as a closed subset. Also we denote by $H(s)$ and $\widetilde{H}(S)$ the inductive limits of $\{H(D) : D \in N(S)\}$ and $\{H(D-S) : D \in N(S)\}$ with respect to the canonical homomorphisms, respectively. $\widetilde{H}(S)$ is regarded as an extension ring of $H(S)$ in a natural manner, and so an $H(S)$-module $M(S)$ may be defined by

$$M(S) \equiv \widetilde{H}(S) \pmod{H(S)}.$$

Each element of $M(S)$ is called a *hyperfunction* on S. By definition, each hyperfunction $g \in M(S)$ is represented by an element of $\widetilde{H}(S)$ and hence by an element $\varphi \in H(D-S)$ with some $D \in N(D)$. We call such φ a *defining function* of g, and write

$$g = [\varphi, D] = [\varphi]$$

or

$$g(x) = [\varphi(z), D]_{z=x} = [\varphi(z)]_{z=x}.$$

Set $I = D \cap \mathbb{R}$, and define $\varphi(x \pm i0) \in M(I)$, to be the boundary values of $\varphi(z) \in H(D-I)$, given by

$$\begin{aligned}\varphi(x+i0) &= [\varepsilon\varphi, D],\\ \varphi(x-i0) &= -[\bar{\varepsilon}\varphi, D]\end{aligned}$$

with

$$\varepsilon(z) = \begin{cases} 1 \ (\operatorname{Im} z > 0) \\ 0 \ (\operatorname{Im} z < 0) \end{cases},$$

$$\bar{\varepsilon}(z) = \varepsilon(-z) = \begin{cases} 0 \ (\operatorname{Im} z > 0) \\ 1 \ (\operatorname{Im} z < 0). \end{cases}$$

Then we have the following representation of $g \in [\varphi, D]$:

$$g(x) = \varphi(x+i0) - \varphi(x-i0).$$

6.2.2.2 *An example due to Hecke*

Hecke defined in [36] the following theta series

$$D(z) = \sum_{0<m+n\sqrt{2}<1+\sqrt{2}} e^{2\pi i z \frac{|m^2-2n^2|}{8}}$$

with $z = x+iy$, $y>0$ and proved that

$$D\left(\frac{az+b}{cz+d}\right) = (cz+d)D(z) + \Psi\left(z, \begin{pmatrix} a & b \\ c & d \end{pmatrix}\right) \quad (c>0)$$

for all $\begin{pmatrix} a & b \\ c & d \end{pmatrix} \in \Gamma$, where Γ is the congruence subgroup of level 8, and $\Psi\left(z, \begin{pmatrix} a & b \\ c & d \end{pmatrix}\right)$ is a polynomial of z. Therefore, $D(z)$ is an *automorphic form of weight* 1 *with period polynomials.* The series $D(z)$ is called the Hecke series. In the following, considering the Hecke series as hyperfunction, we shall prove the transformation formula

$$D\left(\frac{ax+b}{cx+d}\right) = |cx+d| D(x)$$

for all $\begin{pmatrix} a & b \\ c & d \end{pmatrix} \in \Gamma$. We call the series $D(x)$ an *automorphic hyperfunction of weight* 1 for Γ.

We treat the following general forms. Let F be a real quadratic field with discriminant D, and $\mathfrak{O}_F$ the ring of integers in F. Let Q be a natural number and denote by $\mathfrak{U}_0$ the group of totally positive units ε of $\mathfrak{O}_F$ such that $\varepsilon \equiv 1 \pmod{Q\sqrt{D}}$. Let $\mathfrak{a}$ be an integral ideal of $\mathfrak{O}_F$, and put $|N(\mathfrak{a})| = A$. We put

$$D^+\left(z;\, \rho,\, \mathfrak{a},\, Q\sqrt{D}\right) = \sum_{\substack{\mu\in\mathfrak{O}_F \\ \mu\equiv\rho\,(\mathrm{mod}\,\mathfrak{a}Q\sqrt{D}) \\ \mu\in\mathfrak{O}_F/\mathfrak{U}_0,\ \mu\mu'>0}} e^{2\pi i z \frac{\mu\mu'}{AQD}},$$

and

$$D^-\left(z;\, \rho,\, \mathfrak{a},\, Q\sqrt{D}\right) = \sum_{\substack{\mu\in\mathfrak{O}_F \\ \mu\equiv\rho\,(\mathrm{mod}\,\mathfrak{a}Q\sqrt{D}) \\ \mu\in\mathfrak{O}_F/\mathfrak{U}_0,\ \mu\mu'<0}} e^{2\pi i \bar{z} \frac{\mu\mu'}{AQD}},$$

$$= D^+\left(-\bar{z};\, \sqrt{D}\rho,\, \sqrt{D}\mathfrak{a},\, Q\sqrt{D}\right).$$

Then we have the following two formulas:

$$D^+\left(-\frac{1}{z};\, \rho,\, \mathfrak{a},\, Q\sqrt{D}\right) + \frac{\ell(Q\sqrt{D})}{\pi Q\sqrt{D}} iz$$

$$= \pm\frac{z}{Q\sqrt{D}} \sum_{\substack{\beta \bmod \mathfrak{a}Q\sqrt{D} \\ \beta\equiv 0\,(\mathrm{mod}\,\mathfrak{a})}} e^{2\pi i \mathrm{tr}\left(\frac{\rho'\beta}{AQD}\right)} D^+\left(z;\, \beta,\, \mathfrak{a},\, Q\sqrt{D}\right)$$

$$+ \frac{1}{\pi Q\sqrt{D}} \sum_{\substack{\beta \bmod \mathfrak{a}Q\sqrt{D} \\ \beta\equiv 0\,(\mathrm{mod}\,\mathfrak{a})}} e^{2\pi i \mathrm{tr}\left(\frac{\rho'\beta}{AQD}\right)} \int_0^\infty \frac{x}{x+iz} D^+\left(ix;\, \beta,\, \mathfrak{a},\, Q\sqrt{D}\right) dx$$

$$- \frac{1}{\pi Q\sqrt{D}} \sum_{\substack{\beta \bmod \mathfrak{a}Q\sqrt{D} \\ \beta\equiv 0\,(\mathrm{mod}\,\mathfrak{a})}} e^{2\pi i \mathrm{tr}\left(\frac{\rho'\beta}{AQD}\right)} \int_0^\infty \frac{x}{x-iz} D^+\left(ix;\, \sqrt{D}\beta,\, \sqrt{D}\mathfrak{a},\, Q\sqrt{D}\right) dx,$$

where $\ell(Q\sqrt{D}) = \log\varepsilon_0$ with $\mathfrak{U}_0 = \langle\varepsilon_0\rangle$ $(\varepsilon_0 > 1)$, and if $z = |z|e^{i\theta}$, then the signature $+1$ or -1 of the above first term is according to positive or negative of $\operatorname{Re} z = |z|\cos\theta$, respectively;

$$D^-\left(-\frac{1}{z};\,\rho,\,\mathfrak{a},\,Q\sqrt{D}\right) - \frac{\ell(Q\sqrt{D})}{\pi Q\sqrt{D}}i\bar{z}$$

$$= \pm\frac{\bar{z}}{Q\sqrt{D}} \sum_{\substack{\beta \bmod \mathfrak{a}Q\sqrt{D} \\ \beta\equiv 0\,(\mathrm{mod}\,\mathfrak{a})}} e^{2\pi i \mathrm{tr}\left(\frac{\rho'\beta}{AQD}\right)} D^-\left(z;\,\beta,\,\mathfrak{a},\,Q\sqrt{D}\right)$$

$$+\frac{1}{\pi Q\sqrt{D}} \sum_{\substack{\beta \bmod \mathfrak{a}Q\sqrt{D} \\ \beta\equiv 0\,(\mathrm{mod}\,\mathfrak{a})}} e^{2\pi i \mathrm{tr}\left(\frac{\rho'\beta}{AQD}\right)} \int_0^\infty \frac{x}{x-i\bar{z}} D^-\left(ix;\,\beta,\,\mathfrak{a},\,Q\sqrt{D}\right)dx$$

$$-\frac{1}{\pi Q\sqrt{D}} \sum_{\substack{\beta \bmod \mathfrak{a}Q\sqrt{D} \\ \beta\equiv 0\,(\mathrm{mod}\,\mathfrak{a})}} e^{2\pi i \mathrm{tr}\left(\frac{\rho'\beta}{AQD}\right)} \int_0^\infty \frac{x}{x+i\bar{z}} D^+\left(ix;\,\sqrt{D}\beta,\,\sqrt{D}\mathfrak{a},\,Q\sqrt{D}\right)dx,$$

where the signature ± 1 of the first term above is the same in the case of D^+.

Now we put

$$D^+\left(x;\,\rho,\,\mathfrak{a},\,Q\sqrt{D}\right) = \sum_{\substack{\mu\in\mathfrak{O}_F \\ \mu\equiv\rho\,(\mathrm{mod}\,\mathfrak{a}Q\sqrt{D}) \\ \mu\in\mathfrak{O}_F/\mathfrak{U}_0,\ \mu\mu'>0}} e^{2\pi i(x+i0)\frac{\mu\mu'}{AQD}},$$

$$D^-\left(x;\,\rho,\,\mathfrak{a},\,Q\sqrt{D}\right) = \sum_{\substack{\mu\in\mathfrak{O}_F \\ \mu\equiv\rho\,(\mathrm{mod}\,\mathfrak{a}Q\sqrt{D}) \\ \mu\in\mathfrak{O}_F/\mathfrak{U}_0,\ \mu\mu'<0}} e^{2\pi i(x-i0)\frac{\mu\mu'}{AQD}},$$

and

$$\begin{aligned} D(x) &= D\left(x;\,\rho,\,\mathfrak{a},\,Q\sqrt{D}\right) \\ &= D^+\left(x;\,\rho,\,\mathfrak{a},\,Q\sqrt{D}\right) + D^-\left(x;\,\rho,\,\mathfrak{a},\,Q\sqrt{D}\right) \\ &= D^+(x) + D^-(x). \end{aligned}$$

Tending $\varepsilon \to +0$ as $z = x + i\varepsilon$, we have the following transformation formulas:

$$D\left(-\frac{1}{x};\,\rho,\,\mathfrak{a},\,Q\sqrt{D}\right) = |x|\sum_{\beta} c_{\rho,\beta} D\left(x;\,\beta,\,\mathfrak{a},\,Q\sqrt{D}\right),$$

where $c_{\rho,\beta} = \frac{1}{Q\sqrt{D}} e^{2\pi i \mathrm{tr}\left(\frac{\rho'\beta}{AQD}\right)}$. Also, we see easily that

$$D(x+k) = e^{2\pi i k \frac{\rho\rho'}{AQD}} D(x).$$

Summing up the above results, we come to the following

$$D\left(\frac{ax+b}{cx+d}\right) = |cx+d| D(x)$$

for all $\begin{pmatrix} a & b \\ c & d \end{pmatrix} \in \Gamma(QD)$. The function $D(x)$ is one of automorphic hyperfunctions of weight 1 for $\Gamma(QD)$.

The above interpretation of the series $D(z)$ from hyperfunction theoretic point of view is due to M. Sato (1964).

Remark 6.1. Maass constructed in [66] the following real analytic function

$$g(z;\rho) = \ell\left(Q\sqrt{D}\right)\delta(\rho)y^{\frac{1}{2}} + \sum_{\substack{\mu \in \mathfrak{O}_F \\ \mu \equiv \rho \, (\mathrm{mod}\, \sqrt{D}) \\ \mu \in \mathfrak{O}_F/\mathfrak{U}_0}} \sqrt{y} K_0\left(\frac{2\pi|\mu\mu'|}{D} y\right) e^{2\pi i \frac{\mu\mu'}{D} x},$$

where

$$\delta(\rho) = \begin{cases} 1, \ \rho \equiv 0 \quad (\mathrm{mod}\ \sqrt{D}), \\ 0, \ \text{otherwise}. \end{cases}$$

It is easy to see that $g(z;\rho)$ belongs to $M_{\Delta}(\frac{1}{4}, \Gamma)$.

Chapter 7

Selberg's eigenvalue conjecture and the Ramanujan-Petersson conjecture

7.1 Five conjectures in arithmetic

7.1.1 *Selberg's eigenvalue conjecture* (C_1)

If Γ is a discrete subgroup of $PSL_2(\mathbb{R})$. Then the Laplace operator on $\Gamma\backslash S$ is simply

$$\Delta = -y^2\left(\frac{\partial^2}{\partial x^2} + \frac{\partial^2}{\partial y^2}\right).$$

This is a symmetric and non-negative operator which has a self-adjoint extension to all of $L^2(\Gamma \setminus S)$. Spectral decomposition of Δ on $L^2(\Gamma \setminus S)$ decomposes $L^2(\Gamma \setminus S)$ to the direct sum of its discrete and continuous spectrum. As usual we write $\lambda = s(1-s)$, $s \in \mathbb{C}$, to denote an eigenvalue for Δ. We note that the continuous part is $s = \frac{1}{2} + ir$, $r \in \mathbb{R}$, giving $\lambda = \frac{1}{4} + r^2$. For the discrete spectrum, we denote the distinct eigenvalues as

$$0 = \lambda_0 < \lambda_1 < \lambda_2 < \cdots.$$

If $0 < \lambda_i < \frac{1}{4}$, we call λ_i exceptional, and such values are finite in number.

As for non-compact Riemann surfaces, one can find examples $\lambda_1(\Gamma\backslash S) < \frac{1}{4}$. But they are not congruence subgroups. In fact, for a congruence subgroup, Selberg made the following conjecture:

Selberg's eigenvalue conjecture (C_1). *There are no exceptional eigenvalues for congruence subgroups, i.e., one has the bound* $\lambda_1 \geqq \frac{1}{4}$.

In other words, the cuspidal spectrum lies on the continuous one. This conjecture is the fundamental unsolved analytic question in modular forms. It has many applications in arithmetic. The conjecture is known to be true for a few groups of small level. In the case of the modular group

$\Gamma = \mathrm{SL}_2(\mathbb{Z})$, Maass and Roelcke have established the somewhat sharper bounds, $\lambda_1 = 91.14\cdots$. But as the level tends to infinity, one can find many eigenvalues near $\frac{1}{4}$.

Selberg backed up conjecture $\mathbf{C}_1$ by proving the remarkable

Theorem 7.1 (Selberg [88]). *For any congruence subgroup Γ we have*

$$\lambda_1 = \lambda_1(\Gamma \setminus S) \geqq \frac{3}{16}.$$

In 1978, Gelbart and Jacquet, using methods very different from Selberg's, showed that one can replace the equality above by an inequality.

As for the Selberg conjecture the best result established so far is the following

Theorem 7.2 (Kim-Sarnak). *We have*

$$\lambda_1 \geqq \frac{1}{4} - \left(\frac{7}{64}\right)^2 = \frac{975}{4096} \cong 0.2380371.$$

7.1.2 *The Sato-Tate conjecture* ($\mathbf{C}_2$)

Let E be an elliptic curve over a number field F. For each prime ideal v of F where E has good reduction, the number of points of E mod v is given by $N(v)+1-a_v$, where $N(v)$ denotes the norm of v and a_v satisfies Hasse's inequality $|a_v| \leqq 2(N(v))^{1/2}$. Thus we can write

$$a_v = 2N(v)^{\frac{1}{2}} \cos\theta_v$$

for a uniquely defined angle θ_v satisfying $0 \leqq \theta_v \leqq \pi$. The *Sato-Tate conjecture* is a statement concerning how the angles θ_v are distributed in the interval $[0, \pi]$ as v varies. In the non-CM case, we know the distribution of the angles are not uniformly distribution mod 1 when $F = \mathbb{Q}$. Sato and Tate predicted another law regarding the distribution of the angles θ_v. Namely, they predict that

$$\#\{v : N(v) \leqq x,\ \theta_v \in (\alpha, \beta)\} \sim \left(\frac{2}{\pi}\int_\alpha^\beta \sin^2\theta d\theta\right)\pi_F(x)$$

as x tends to infinity, where $\pi_F(x)$ is the number of prime ideals of F whose norm is less than x.

Langlands outlined an approach to the Sato-Tate conjecture using the theory of automorphic forms. Thus the conjecture of Sato-Tate was applicable in a larger context to modular forms or more generally, to automorphic forms on GL_2.

For example, one could take the *Ramanujan τ function* attached to the unique newform of weight 12 and level 1, and write

$$\tau(p) = 2p^{11/2} \cos \theta_p.$$

Then it is conjectured that the θ_p s are uniformly distributed in $[0, \pi]$ with respect to the measure

$$\frac{2}{\pi} \sin^2 \theta.$$

This was first proposed by M. Sato and Serre, and is called the Sato-Tate conjecture because of the close analogy it bears to a conjecture of Sato and Tate on elliptic curves. If for each prime p we associate a conjugacy class X_p in a compact group G, then given an irreducible representation ρ of G, define the L-series

$$L(s, \rho) = \prod_p \det\left(1 - \rho(X_p)p^{-s}\right)^{-1}.$$

Serre proved that if for every non-trivial irreducible representation ρ of G, it holds that

(a) $L(s, \rho)$ has an analytic continuation to $\operatorname{Re} s \geqq 1$ and
(b) $L(1 + it, \rho) \neq 0$ for $t \in \mathbb{R}$,

then the X_p s are uniformly distributed in G with respect to the *Haar measure* of G ([92]). In this case, let $G = \mathrm{SU}_2(\mathbb{C})$, and the conjugacy classes of G are parametrized by $0 \leq \theta \leqq \pi$ and each class has a representation of the form

$$\begin{pmatrix} e^{i\theta} & \\ & e^{-i\theta} \end{pmatrix}.$$

The irreducible representations are given by the sequence $\rho_0, \rho_1, \ldots$ where ρ_n has character χ_n given by

$$\chi_n \begin{pmatrix} e^{i\theta} & \\ & e^{-i\theta} \end{pmatrix} = \frac{\sin(n+1)\theta}{\sin \theta}.$$

Therefore, Serre's theorem states that if for all $n \geqq 1$,

(a) $L(s, \rho_n)$ has an analytic continuation to $\operatorname{Re} s \geqq 1$ and
(b) $L(1 + it, \rho_n) \neq 0$ for $t \in \mathbb{R}$,

then the Sato-Tate conjecture for $\tau(p)$ is true. K. Murty subsequently showed that (a) alone suffices to imply the Sato-Tate conjecture. That is, the non-vanishing condition turns out to be consequence of analytic continuation to the line $\operatorname{Re} s = 1$. For instance, if $n = 0$, then $L(s, \rho_0) = \zeta(s)$, the *Riemann zeta-function.* If $n = 1$, then

$$L(s, \rho_1) = \sum_{n=1}^{\infty} \frac{\tau(n)}{n^{s+\frac{11}{2}}}$$

is the series introduced by Ramanujan and attached by Hecke to a classical cusp form of weight 2. For $n = 2$, *Rankin-Selberg theory* allows one to deduce that $L(s, \rho_2)$ extends to an entire function for $\operatorname{Re} s \geqq 1$. In a recent work, Kim and Shahidi showed that $L(s, \rho_3)$ extends to an entire function and later, Kim showed the same for $L(s, \rho_4)$. For the cases $5 \leq n \leqq 9$, Kim and Shahidi have shown that ρ_n extends to a meromorphic function for all $s \in \mathbb{C}$ which is regular for $\operatorname{Re} s \geqq 1$, except in the case of $m = 9$, $L(s, \rho_9)$ may have a pole at $s = 1$.

In the following, we treat the general formulation of the Sato-Tate conjecture which is due to Langlands.

Let X be a compact topological space and denoted by $C(X)$ *Banach space* of continuous complex valued functions on X defined by the sup norm

$$\|f\| = \sup_{x \in X} |f(x)|.$$

Given $x \in X$, let δ_x be the Dirac measure defined by $\delta_x(f) = f(x)$, $f \in C(X)$. If $\{x_n\}_{n \geqq 1}$ is a sequence of points in X, we set

$$\mu_n = \frac{\delta_{x_1} + \cdots + \delta_{x_n}}{n}$$

and let μ be a continuous linear form on $C(X)$. Then the sequence $\{x_n\}_{n \geqq 1}$ is *μ-equidistribution* or *μ-uniformly distributed*, if $\mu_n \to \mu$ weakly, i.e., $\mu_n(f) \to \mu(f)$ as $n \to \infty$ for every $f \in C(X)$.

Let G be a compact group and take X to be the space of conjugacy classes of G. More precisely $X = G/\sim$, where $x \sim y$ if and only if there exists g such that $x = g^{-1}yg$. Fix a Haar measure μ on G and use the same notation to define its image in X. Let χ be an irreducible character of G and set

$$\mu(x) = \int \chi(g) d\mu(g).$$

Then, the sequence $\{x_n\}_{n \geqq 1} \subset X$ is μ-equidistributed if and only if for every irreducible character χ of G, it holds that

$$\lim_{n \to \infty} \frac{1}{n} \sum_{i=1}^{n} \chi(x_i) = \mu(x).$$

Assume that μ is so normalized that $\mu(\mathbf{1}) = 1$. Then by the orthogonality of characters we conclude that

Theorem 7.3. *The sequence* $\{x_n\}_{n \geqq 1} \subset X$ *is* μ*-equidistributed if and only if*

$$\lim_{n\to\infty} \frac{1}{n} \sum_{i=1}^{n} \chi(x_i) = 0$$

for every $\chi \neq \mathbf{1}$.

Now take $G = \mathrm{SU}_2(\mathbb{C})$. It can be identified with S^3, 3-sphere of radius 1. The action of G on itself by conjugation transfers to an action of S^3 by rotation. The Haar measure on G is then a rotational invariant measure on S^3. Every conjugacy class of G contains a unique element of the form

$$t_\theta = \begin{pmatrix} e^{i\theta} & \\ & e^{-i\theta} \end{pmatrix} \quad (0 \leqq \theta \leqq \pi)$$

which by means of the above identification is isomorphic to S^2_θ, 2-sphere of radius $|\sin\theta|$. Therefore, the normalized measure on S^3, giving measure 1 for G, induces the measure $\frac{2}{\pi}\sin^2\theta d\theta$ on conjugacy classes of G.

Let $\pi = \otimes_v \pi_v$ be a cuspidal representation of $\mathrm{GL}_2(\mathbb{A}_F)$. Suppose v is an archimedean place of F. Then π_v is parametrized by a 2-dimensional representation ρ_v of the Weil group W_{F_v} of $\bar{F}_v/F_v$. Suppose either

a) π is monomial, i.e., $\pi \cong \pi \otimes \eta$ for some nontrivial character η of $F^\times \backslash \mathbb{A}_F^\times$.

or,

b) For every archimedean place v of F, the representation ρ_v parametrizing π_v is of Galois type, i.e., factors through the Galois group of $\bar{F}_v/F_v$.

We want to avoid π being a twist of either of these types, i.e., that it be parametrized by the 2-dimensional representation of the Weil group W_F of $\bar{F}/F$. Clearly those of type a) comprise all of those that are parametrized by 2-dimensional nonprimitive irreducible representations of W_F. While those which are twists of representations of type b) are parametrized, at least conjecturally, by the remaining irreducible 2-dimensional representations of W_F. We therefore define: A cuspidal representation π of $\mathrm{GL}_2(\mathbb{A}_F)$ is called *genuine* if it is not a twist of a representation of either type a) or type b).

When $F = \mathbb{Q}$ and π comes from a Maass form, this is attained by assuming that the defining $s \neq \frac{1}{4}$. On the other hand, no holomorphic modular form of even weight will be of Galois type. Assume that π has a trivial central character and that it satisfies the Ramanujan-Petersson conjecture. Then at each v for which π_v is of class one, the corresponding conjugacy class in $\mathrm{GL}_2(\mathbb{C})$ is given by

$$t_v = t_{\theta_v} = \begin{pmatrix} e^{i\theta_v} & \\ & e^{-i\theta_v} \end{pmatrix} \quad (0 \leqq \theta_v \leqq \pi).$$

The Sato-Tate conjecture ($\mathbf{C}_2$). *Assume that π is genuine. Then the angles θ_v are equidistributed with respect to the natural measure* $\frac{2}{\pi}\sin^2\theta d\theta$.

Recently, the original Sato-Tate conjecture was proved by R. Taylor under a mild technical hypothesis which well undoubtedly be removed: *Automorphy for some ℓ-adic lifts of automorphic mod ℓ Galois representations II*, Publ. Math. Inst. Hautes Etudes Sci. **108** (2008), 183-239.

7.1.3 *The Ramanujan-Petersson conjecture* ($\mathbf{C}_3$)

Let Γ be a discrete subgroup of $\mathrm{PSL}_2(\mathbb{R})$. To study the eigenvalue of the Laplace operator Δ on $\Gamma \setminus S$, one must consider the eigenfunctions. These are functions on $\Gamma \setminus S$ and those for eigenvalue $\lambda = s(1-s)$ can be written as

$$f(x,y) = a_0 y^s + a_0' y^{1-s} + \sum_{n \neq 0} a_n \sqrt{y} K_{s-\frac{1}{2}}(2\pi|n|y) e^{2\pi i n x},$$

where $K_\nu(z)$ is the *Whittaker-Bessel function* bounded at infinity, i.e., the solution to the differential equation

$$t^2 K_\nu'' + t K_\nu' - (t^2 + \nu^2) K_\nu = 0$$

satisfying

$$K_\nu(z) \sim \sqrt{\frac{\pi}{2t}} e^{-t}$$

as t goes to $+\infty$. The complex numbers a_n are the corresponding Fourier coefficients. Then we can state the following conjecture.

The Ramanujan-Petersson conjecture ($\mathbf{C}_3$). *For every Maass cusp form f which is an eigenfunction for all the Hecke operators with $a_1 = 1$, and for every prime number p,*

$$|a_p| \leqq 2.$$

There is an analogue of the Ramanujan-Petersson conjecture for normalized holomorphic cuspidal eigenforms of weight k, stating $|a_n| \leqq 2p^{\frac{k-1}{2}}$. This was proved by Deligne in 1973, as a consequence of his proof of Weil's conjecture.

7.1.4 *Linnik-Selberg's conjecture* ($\mathbf{C}_4$)

The sum

$$S(r,n,c) = \sum_{\substack{d \pmod c \\ ad \equiv 1 \pmod c}} e^{\frac{2\pi i}{c}(nd+ar)}$$

is called a *Kloosterman sum.* Regarding this, Linnik and Selberg independently conjectured that

Linnik-Selberg's conjecture ($\mathbf{C}_4$). *For* $x \geqq \gcd(r,n)^{\frac{1}{2}+\varepsilon}$ *and for any* $\varepsilon > 0$,

$$\sum_{c \leqq x} \frac{S(r,n,c)}{c} = O(x^{\varepsilon}).$$

Selberg's original motivation for this conjecture was that it could yield a proof of the Ramanujan-Petersson conjecture in the holomorphic case (for Maass forms as well), but he did not indicate a proof.

7.1.5 *The Gauss-Hasse conjecture* ($\mathbf{C}_5$)

The Gauss-Hasse conjecture ($\mathbf{C}_5$). *There exists an infinite number of real quadratic fields with class number one.*

Hasse throughout his life advocated this statement of Gauss both in private talks and his books. Hence we think it proper to combine their names in the title of the conjecture.

7.2 Some relations between the five conjectures

7.2.1 *Conjectures* $\mathbf{C}_1$ *and* $\mathbf{C}_3$

In 1961, Satake interpreted Selberg's eigenvalue conjecture as an analogue of the Ramanujan-Petersson conjecture for the 'infinite place,' so today the conjectures $\mathbf{C}_1$ and $\mathbf{C}_3$ lie within the scope of the program of Langlands.

We discuss a simple example of Hecke in which the ideas are self evident. Let $k = \mathbb{Q}(\sqrt{d})$ be a real quadratic field with discriminant d. Let f be a positive integer and denote by R_f the class ring of discriminant $d_f = df^2$ and conductor f. Let $\varepsilon_f > 1$ be a generator of the group of units $R_f^{\times}$. Let $\mathbb{T} = \{z \in \mathbb{C} : z\bar{z} = 1\}$ be the circle group with the topology and let

$$\Psi : k_{\mathbb{A}}^{\times}/k^{\times} \to \mathbb{T}$$

be the Hecke character which is unramified at all primes v including the archimedean one, and whose infinite component has the value

$$\Psi_\infty((\alpha R_f)) = \left|\frac{\alpha}{\alpha'}\right|^{\frac{i\pi}{\log\varepsilon_f}}$$

at the principal idele generated by the element α. For each integer n we have an associated Hecke L-function

$$\begin{aligned} L(s,\Psi^n) &= \Gamma_{\mathbb{R}}\left(s - \frac{i\pi n}{\log\varepsilon_f}\right)\Gamma_{\mathbb{R}}\left(s + \frac{i\pi n}{\log\varepsilon_f}\right)\prod_{v\in S_f}\frac{1}{1-\Psi_v^n(\pi_v)Nv^{-s}} \\ &= \Gamma_{\mathbb{R}}\left(s - \frac{i\pi n}{\log\varepsilon_f}\right)\Gamma_{\mathbb{R}}\left(s + \frac{i\pi n}{\log\varepsilon_f}\right)\prod_{v\in S_f}\frac{1}{1-a_p p^s + \left(\frac{d_f}{p}\right)p^{-2s}}, \end{aligned}$$

where $\Gamma_{\mathbb{R}}(s) = \pi^{-\frac{s}{2}}\Gamma\left(\frac{s}{2}\right)$, π_v is a local uniformizing parameter for each finite prime v of k. We note that the finite part is a quadratic Euler product and hence $L(s,\Psi^n)$ is the L-function of a cuspidal automorphic representation $\pi(\Psi) = \otimes_p\pi_p$ on the group $\mathrm{GL}_2(\mathbb{Q}_{\mathbb{A}})$, and we have in particular that $L(s,\Psi^n)$ is holomorphic for $\mathrm{Re}\, s \geqq 1$, except for a simple pole at $s = 1$ when $n = 0$. Then we have the following three properties.

(1) (Ramanujan-Petersson conjecture)

$$|a_p| \leqq 2$$

for all primes p.

(2) (Ramanujan-Petersson conjecture at infinity) The eigenvalue of the Laplacian operator satisfies

$$\lambda_\infty(\pi(\Psi)) \geqq \frac{1}{4},$$

where $\lambda_\infty = \dfrac{1-s^2}{4}$ with $s = \dfrac{i\pi n}{\log\varepsilon_f}$.

(3) $L(1+it,\Psi^n) \neq 0$ for all $t \in \mathbb{R}$.

7.2.2 *Conjectures* $\mathbf{C}_1$ *and* $\mathbf{C}_5$

Let p be a prime number such that $p \equiv 1 \pmod 4$. Then Takhtajan and Vinogradov proved in [101] that the point $\lambda = \frac{1}{4}$ belongs to the spectrum of Δ on the space $S_\Delta(\lambda, \Gamma_0(p), \chi_p)$ whenever the class number of the real quadratic field $\mathbb{Q}(\sqrt{p})$ is greater than one. Here χ_p is a quadratic character given by

$$\chi_p(a) = \left(\frac{a}{p}\right).$$

Moreover, from the spectral point of view it seems plausible that for 'almost all' D the discrete spectrum of the Laplacian Δ in the Hilbert space $L^2(\Gamma_0(D)\backslash S, \chi_D)$ lies to the right of the point $\lambda = \frac{1}{4}$. Certainly the Gauss-Hasse conjecture follows from this result.

7.2.3 *Conjectures* $\mathbf{C}_3$ *and* $\mathbf{C}_4$

Let F be a number field and denote by $\mathbb{A}_F$ its ring of adeles. Let π be a cuspidal representation of $\mathrm{GL}_2(\mathbb{A}_F)$, that is an irreducible constituent of the space of cuspidal integrable functions φ on $Z(\mathbb{A}_F)\mathrm{GL}_2(F) \backslash \mathrm{GL}_2(\mathbb{A}_F)$ satisfying $\varphi(zg) = \omega(z)\varphi(g)$, where ω is the central character of π and $z \in Z(\mathbb{A}_F) = \mathbb{A}_F^\times$, the center of $\mathrm{GL}_2(\mathbb{A}_F)$. Here cuspidal means

$$\int_{F\backslash\mathbb{A}_\mathbb{Q}} \varphi\left(\begin{pmatrix} 1 & x \\ 0 & 1 \end{pmatrix} g\right) dx = 0$$

for almost all $g \in \mathrm{GL}_2(\mathbb{A}_F)$. When $F = \mathbb{Q}$, they account for all of the classical modular cusp forms and Maass cusp forms. Therefore from this, holomorphic forms and real analytic forms inhabit the same stage from the representational point of view.

The Ramanujan-Petersson conjecture for holomorphic case is essentially identical to the conjecture $\mathbf{C}_4$ that was mentioned in Linnik ([16]).

The *Poincaré series* for $\Gamma = \mathrm{SL}_2(\mathbb{Z})$ are defined by

$$G_r(z) = \frac{1}{2} \sum_{(c,d)=1} (cz+d)^{-k} e^{2\pi i r \frac{az+b}{cz+d}},$$

where a, b are any integers such that $ad - bc = 1$. For every $r \geqq 1$, the function $G_r(z)$ is cusp form of weight k for Γ. Let f be any cusp form of weight k, and the Fourier expansion of $f(z)$:

$$f(z) = \sum_{n=1}^{\infty} a_n e^{2\pi i n x} \cdot e^{-2\pi n y}.$$

Then the inner product (f, G_r) is the following

$$(f, G_r) = \frac{\Gamma(k-1)}{(4\pi r)^{k-1}} a_r.$$

From this *inner product formula*, we have that every cusp form is a finite linear combination of Poincaré series.

Expanding $G_r(z)$ in Fourier series,

$$G_r(z) = \sum_{n=1}^{\infty} b_{rn} e^{2\pi i n z}.$$

Then,

$$\begin{aligned} b_{rn} &= \int_0^1 G_r(z) e^{-2\pi i x} dx \\ &= \frac{1}{2} \sum_{(c,d)=1} \int_0^1 (cx+d)^{-k} e^{2\pi i r\left(\frac{ax+b}{cx+d}\right) - 2\pi i n x} dx \\ &= \frac{1}{2} \sum_{c \neq 0} \frac{1}{c} \sum_{\substack{d \,(\mathrm{mod}\, c) \\ ad \equiv 1 \,(\mathrm{mod}\, c)}} e^{\frac{2\pi i}{c}(nd+ar)} \int_{-\infty}^{\infty} t^{-k} e^{-\frac{2\pi i}{c}\left(\frac{r}{t}+nt\right)} dt; \end{aligned}$$

$$\int_{-\infty+ci}^{\infty+ci} t^{-k} \exp\left(-\frac{2\pi i}{c}\left(\frac{r}{t}+nt\right)\right) dt = 2\pi \left(\frac{n}{r}\right)^{\frac{k-1}{2}} J_{k-1}\left(\frac{4\pi\sqrt{rn}}{c}\right),$$

where $J_k(z)$ is a Bessel function defined by

$$J_k(z) = \frac{\left(\frac{z}{2}\right)^k}{\sqrt{\pi}\,\Gamma\left(k+\frac{1}{2}\right)} \int_0^{\pi} \sin^{2k}\theta \cos(z\cos\theta) d\theta.$$

Finally we obtain the formula due to Petersson:

$$b_{rn} = \left(\frac{n}{r}\right)^{\frac{k-1}{2}} \left\{ \delta_{rn} + \pi \sum_{c=1}^{\infty} \frac{S(r,n,c)}{c} J_{k-1}\left(\frac{4\pi\sqrt{rn}}{c}\right) \right\},$$

where δ_{rn} denotes the Kronecker delta function.

We have already noted that the Poincaré series span the space of cusp forms. Therefore, to prove the Ramanujan-Petersson conjecture for the holomorphic case, it suffices to show that

$$b_{rn} = O\left(n^{\frac{k-1}{2}+\varepsilon}\right)$$

for every r. This is tantamount to showing that the expression in parentheses in the above sum is $O(n^{\varepsilon})$. This is implied by the Linnik-Selberg conjecture $\mathbf{C}_4$. The argument was carried out by M. R. Murty ([16]).

As mentioned above, the Linnik-Selberg conjecture implies the Ramanujan-Petersson conjecture for the holomorphic case (for the full modular group $\mathrm{SL}_2(\mathbb{Z})$).

7.2.4 *Conjectures* $\mathbf{C}_2$ *and* $\mathbf{C}_3$

Let S be a finite set of places of the number field. Then the conjecture $\mathbf{C}_2$ and the conjecture $\mathbf{C}_3$ with respect to S can both be proved by considering the same analytic properties of the *symmetric power L-functions* for GL_2 (cf., Shahidi [92]).

Chapter 8

Indefinite theta series

8.1 Indefinite quadratic forms and indefinite theta series

8.1.1 *Hecke's indefinite theta series*

In this subsection we shall review the definition and basic properties of the *indefinite theta series* that were introduced in Section 3.1.

Let F be a real quadratic field with discriminant D, and $\mathfrak{O}_F$ the ring of all integers in F. Let Q be a natural number and denote by $\mathfrak{U}_0$ the group of totally positive units ε of $\mathfrak{O}_F$ such that $\varepsilon \equiv 1 \pmod{Q\sqrt{D}}$. Let $\mathfrak{a}$ be an integral ideal of $\mathfrak{O}_F$, and put $|N(\mathfrak{a})| = A$. Then the Hecke theta series for $\mathfrak{a}$ is defined by

$$\vartheta_\kappa\left(\tau;\, \rho,\, \mathfrak{a},\, Q\sqrt{D}\right) = \sum_{\substack{\mu \in \mathfrak{O}_F \\ \mu \equiv \rho \,(\mathrm{mod}\, \mathfrak{a}Q\sqrt{D}) \\ \mu \in \mathfrak{O}_F/\mathfrak{U}_0,\, \mu\mu'\kappa > 0}} \operatorname{sgn}(\mu) \cdot q^{\frac{|\mu\mu'|}{AQD}},$$

where $\kappa = \pm 1$, $\rho \in \mathfrak{a}$, $\mathrm{Im}\,\tau > 0$ and $q = e^{2\pi i \tau}$. This is a holomorphic function of τ and satisfies

$$\vartheta_\pm\left(\frac{a\tau + b}{c\tau + d};\, \rho,\, \mathfrak{a},\, Q\sqrt{D}\right) = \left(\frac{D}{|d|}\right) e^{\mp 2\pi i \frac{ab\rho\rho'}{AQD}} (c\tau + d) \vartheta_\pm\left(\tau;\, a\rho,\, \mathfrak{a},\, Q\sqrt{D}\right),$$

for all $\begin{pmatrix} a & b \\ c & d \end{pmatrix} \in \Gamma_0(QD)$. Therefore $\vartheta_\pm$ is the cusp form of weight 1 for a certain congruence subgroup of level QD under the condition $\vartheta_\pm \not\equiv 0$. If in particular $\mathfrak{a} = \mathfrak{O}_F$, we put

$$\vartheta_\pm\left(\tau;\, \rho,\, Q\sqrt{D}\right) = \vartheta_\pm\left(\tau;\, \rho,\, \mathfrak{O}_F,\, Q\sqrt{D}\right).$$

It is hard to judge whether $\vartheta_\pm$ vanishes identically or not.

8.1.2 *Polishchuk's indefinite theta series*

Let

$$Q = am^2 + 2bmn + cn^2$$

be a $\mathbb{Q}$-valued indefinite quadratic form on $\mathbb{Z}^2$ which is positive on the cone $mn \geqq 0$, i.e., a, b and c are positive. Let $f(m,n)$ be a doubly periodic complex valued function on $\mathbb{Z}^2$. Therefore,

$$f(m+N,n) = f(m,n+N) = f(m,n)$$

for some N. Let us extend the function $f(m,n)$ from $\mathbb{Z}^2$ to $\mathbb{Q}^2$ by zero. Then we impose the following condition on Q and f such that

$$f(A\mathbf{x}) = f(B\mathbf{x}) = -f(\mathbf{x}) \tag{8.1}$$

for every $\mathbf{x} \in \mathbb{Q}^2$, where

$$A = \begin{pmatrix} -1 & -\frac{2b}{a} \\ 0 & 1 \end{pmatrix},$$

$$B = \begin{pmatrix} 1 & 0 \\ -\frac{2b}{c} & -1 \end{pmatrix}.$$

In addition assume that $\frac{1}{2}Q$ takes integer values on the support of f. Then Polishchuk defined the following series in [76]:

$$\Theta_{Q,f} = \left\{ \sum_{m\geqq 0,\, n\geqq 0} f(m,n) - \sum_{m<0,\, n<0} f(m,n) \right\} q^{Q(m,n)}.$$

Theorem 8.1 (Polishchuk). *The notation and assumptions being as above, the series $\Theta_{Q,f}$ is a modular form of weight 1 for a certain congruence subgroup, and the space of modular forms of weight 1 spanned by these series coincides with the space generated by Hecke's indefinite theta series.*

Proof. Let $S \subset \mathbb{Z}^2 \subset \mathbb{Q}^2$ be the support of f, and let $\Lambda = \{x \in \mathbb{Q}^2 : S + x = S\}$. Since f is doubly periodic, Λ is a sublattice of $\mathbb{Z}^2$. On the other hand, both of the operators A and B preserve Λ. Therefore $\mathrm{tr}(AB) = -2 + \frac{4b^2}{ac}$ is an integer, i.e., $\frac{4b^2}{ac}$ is an integer. Making the change of variables of the form $m = \frac{m'}{m_0}$, $n = \frac{n'}{n_0}$, where m_0 and n_0 are positive integers such that $\frac{m_0}{n_0} = \frac{a}{2b}$, we can assume that $a = 2b$. Then both A and B are integer matrices. In particular, we can consider these as acting on $(\mathbb{Z}/N\mathbb{Z})^2$. Let us denote by G_N the subgroup of $\mathrm{GL}_2(\mathbb{Z}/N\mathbb{Z})$ generated by A and B. Then G_N is a dihedral group. Now the space of functions f on $(\mathbb{Z}/N\mathbb{Z})^2$ satisfying the condition (8.1) is spanned by functions supported on orbits of G_N that satisfy (8.1). Let $O \subset (\mathbb{Z}/n\mathbb{Z})^2$ be an orbit of G_N, and f be a function on O satisfying (8.1). In order for f to be non-zero, the orbit O should satisfy the condition that for every $x \in O$ one has $A\mathbf{x} \neq \mathbf{x}$, $B\mathbf{x} \neq \mathbf{x}$. Let us call such an orbit admissible. Conversely, for every admissible orbit O there is a unique function f on O satisfying (8.1). Indeed, let $\chi : G_N \to \{\pm 1\}$ be the character defined by $\chi(A) = \chi(B) = -1$. Then the orbit is admissible if and only if χ is trivial on the stabilizer subgroup of a point in O. Thus, for every admissible orbit $O = Gx$ we can define the function $f_0(gx) = \chi(g)$ up to a sign. The function f_0 on O does not depend on x. So, we can assume that f is a doubly periodic function on $\mathbb{Z}^2$ with values ± 1 satisfying the condition (8.1). Put $S = S_1 \cup S_{-1}$, where $S_1 = f^{-1}(1)$, $S_{-1} = f^{-1}(-1)$. Then $AS_1 = BS_1 = S_{-1}$.

Let $D = b^2 - 4ac$ (> 0) be not a complete square and K be the real quadratic field associated with the form Q. We have

$$Q(m,n) = \frac{1}{c} N\left(bm + nc + m\sqrt{D}\right),$$

and consider $\mathbb{Z}^2$ as a lattice in K via the map $(m,n) \mapsto \left(bm + nc + m\sqrt{D}\right)$. For two non-zero elements $k_1, k_2 \in K$, we denote $\langle k_1, k_2\rangle = \mathbb{Q}_{>0}k_1 + \mathbb{Q}_{>0}k_2$, $[k_1,k_2] = \mathbb{Q}_{\geq 0}k_1 + \mathbb{Q}_{\geq 0}k_2$ and $\langle k_1, k_2] = \mathbb{Q}_{>0}k_1 + \mathbb{Q}_{\geq 0}k_2$. Then we write

$$\Theta_{Q,f} = \sum_{\lambda \in S_1 \cap [1, b+\sqrt{D}]} q^{\frac{N(\lambda)}{c}} - \sum_{\lambda \in S_1 \cap \langle -1, -b-\sqrt{D}\rangle} q^{\frac{N(\lambda)}{c}} - \sum_{\lambda \in S_{-1} \cap [1, b+\sqrt{D}]} q^{\frac{N(\lambda)}{c}}$$

$$+ \sum_{\lambda \in S_{-1} \cap \langle 1, b+\sqrt{D}\rangle} q^{\frac{N(\lambda)}{c}}.$$

Let us extend the operators A and B from the lattice to K by $\mathbb{Q}$-linearity, so that we have $B(1) = -1$, $B(b+\sqrt{D}) = -b+\sqrt{D}$. Hence, making the change of variables $\lambda \mapsto B(\lambda)$ in the last two sums we have

$$\Theta_{Q,f} = \sum_{\lambda \in S_1 \cap \langle b-\sqrt{D}, b+\sqrt{D}]} q^{\frac{N(\lambda)}{c}} - \sum_{\lambda \in S_1 \cap [-b+\sqrt{D}, -b-\sqrt{D}\rangle} q^{\frac{N(\lambda)}{c}}.$$

It is easy to see that the operator $AB : K \to K$ coincides with multiplication by the element $\frac{b+\sqrt{D}}{b-\sqrt{D}}$ of norm 1. Therefore, we have

$$\Theta_{Q,f} = \sum_{\lambda \in S_1 \cap C/G} \operatorname{sgn}(\lambda) q^{\frac{N(\lambda)}{c}},$$

where C is the set of elements with positive norm in K, and G is the infinite cyclic group generated by AB. Note that the set S_1 is a union of a finite number of cosets $\{\Lambda_1 + x_i : i = 1, 2, \ldots, s\}$ for the lattice $\Lambda_1 = \{x \in K : S_1 + x = S_1\}$, and since Λ_1 is preserved by the action of G, there is a subgroup of finite index $G_0 \subset G$ preserving each of these cosets. Hence we have

$$[G : G_0]\Theta_{Q,f} = \sum_{i=1}^{s} \sum_{\lambda \in (\Lambda_1 + x_i) \cap C/G_0} \operatorname{sgn}(\lambda) q^{\frac{N(\lambda)}{c}}. \tag{8.2}$$

Here let us recall the definition of Hecke's indefinite theta series from Polishchuk's point of view. Under the same notation as above, the cone C is a union of two components and we define the sign : $C \to \{\pm 1\}$ which assigns a value of 1 (resp. -1) on totally positive (resp. negative) elements in K. Let us denote by $U_+(K)$ the subgroup of the multiplicative group $K^\times$ consisting of totally positive elements $k \in K^\times$ with norm 1. The group of $\mathbb{Q}$-linear automorphisms of K preserving norm decomposes as follows:

$$\operatorname{Aut}_{\mathbb{Q}}(K, \text{norm}) = \pm\text{id.} \times U_+(K) \times \operatorname{Gal}(K/\mathbb{Q})$$

where $U_+(K)$ acts on K by multiplication. Let $\Lambda \subset K$ be a lattice, $\lambda + c$ be a coset for this lattice. Then Hecke's indefinite theta series is

$$\Theta_{\Lambda,c} = \sum_{\lambda \in (\Lambda + c) \cap C/G} \operatorname{sgn}(\lambda) q^{d \cdot N(\lambda)},$$

where G is the subgroup in $U_+(K)$ consisting of the elements preserving $\Lambda + c$, d is a positive rational number such that $d \cdot$ norm takes integer values on $\Lambda + c$.

Now each of the terms in (8.2) is a scalar multiple of Hecke's series.

Conversely, assume that we are given $\Lambda \subset K$ in a real quadratic field and a coset $\Lambda + c$. Let $G \subset U_+(K)$ be the subgroup preserving $\Lambda + c$. Recall that G is an infinite cyclic group, and let ε be a generator of G. Let us define the $\mathbb{Q}$-linear operators A and B on K as follows: $B(x) = -\bar{x}$, $A(x) = -\varepsilon\bar{x}$ with the conjugate $\bar{x}$ of x. We know that $A^2 = B^2 = 1$, and $\det A = \det B = -1$. Let $k \in K$ be an eigenvector for A with eigenvalue -1, so that $\varepsilon\bar{k} = k$. Changing k by $-k$ if necessary we can assume that k is totally positive. Then we have

$$\Theta_{\Lambda,c} = \sum_{\lambda \in (\Lambda + c) \cap C/G} \operatorname{sgn}(\lambda) q^{dN(\lambda)}$$

$$= \sum_{\lambda \in (\Lambda + c) \cap [k, \bar{k})} q^{dN(\lambda)} - \sum_{\lambda \in (\Lambda + c) \cap (-k, -\bar{k}]} q^{dN(\lambda)}.$$

We have $1 \in \langle k, \bar{k}\rangle$ since k is totally positive. Hence we can split each of the above sums into two according to decompositions $[k, \bar{k}\rangle = [k, 1] \cup \langle 1, \bar{k}\rangle$, $\langle -k, -\bar{k}] = \langle -k, -1\rangle \cup [-1, -\bar{k}]$. Making the change of variable $\lambda \to B(\lambda)$ in the sums over $\langle 1, \bar{k}\rangle$ and $[-1, -\bar{k}]$, we have

$$\Theta_{\Lambda,c} = \sum_{\lambda \in S \cap ([1,k] \cup \langle -k,-1\rangle)} f(\lambda)\mathrm{sgn}(\lambda) q^{d \cdot N(\lambda)},$$

where $S = (\Lambda + c) \cup B(\Lambda + c)$, the function f supported on S defined by $f(x) = \delta_{\Lambda+c}(x) - \delta_{B(\Lambda+c)}(x)$ with the characteristic function δ_I of a set I. Note that since the operator AB preserves $\Lambda + c$, and $(AB)B = B(AB)^{-1}$, the operator also preserves $B(\Lambda + c)$, hence $f(AB\mathbf{x}) = f(\mathbf{x})$. On the other hand, by definition $f(B\mathbf{x}) = -f(\mathbf{x})$, and also $f(A\mathbf{x}) = -f(\mathbf{x})$. Now taking the coordinates with respect to the basis $(1, k)$ as variables of summation the above series can rewrite in the form

$$\Theta_{\Lambda,c} = \sum_{\substack{(m,n)\in S \\ m \geqq 0,\, n \geqq 0}} f(m,n) q^{Q(m,n)} - \sum_{\substack{(m,n)\in S \\ m<0,\, n<0}} f(m,n) q^{Q(m,n)},$$

where S is a finite union of cosets with respect to some $\mathbb{Z}$-lattice in $\mathbb{Q}^2$. In order to rewrite this series in the required form, it only remains to change the variables (m, n) to (Mm, Mn) where $MS \subset \mathbb{Z}^2$. □

It is important open problem to formulate the necessary and sufficient conditions for the series $\Theta_{\Lambda,c}$ to be zero. In other words, the problem is to describe all linear relations between such series for some basis in the space of functions f satisfying the assumptions of Theorem 8.1. Some non-vanishing results were proved by Polishchuk using *homological mirror symmetry* ([77]).

Chapter 9

Hilbert modular forms of weight 1

9.1 Hilbert modular forms

9.1.1 *Hilbert modular groups*

Let K be an algebraic number field and put $n = [K : \mathbb{Q}]$. Such a number field admits n different imbeddings into $\mathbb{C}$: For $1 \leqq j \leqq n$,

$$\begin{array}{ccc} K & \longrightarrow & \mathbb{C} \\ \Downarrow & & \Downarrow \\ a & \longmapsto & a^{(j)}. \end{array}$$

We assume that K is *totally real*, i.e., the image $K^{(j)}$ of each of the n imbeddings is contained in $\mathbb{R}$:

$$\begin{array}{ccc} K & \xrightarrow{\sim} & K^{(j)} \subset \mathbb{R} \\ \Downarrow & & \Downarrow \\ a & \longmapsto & a^{(j)}. \end{array}$$

Sometimes it is useful to identify a and the vector $(a^{(1)}, \ldots, a^{(n)})$. If we attach to the matrix

$$M = \begin{pmatrix} a & b \\ c & d \end{pmatrix} \in \mathrm{GL}_2(K)$$

the tuple

$$(M^{(1)}, \ldots, M^{(n)}),$$
$$M^{(j)} = \begin{pmatrix} a^{(j)} & b^{(j)} \\ c^{(j)} & d^{(j)} \end{pmatrix} \quad (j = 1, \ldots, n),$$

we obtain an imbedding of groups

$$\mathrm{GL}_2(K) \hookrightarrow \mathrm{GL}_2(\mathbb{R})^n.$$

The *Hilbert modular group* of totally real number field K is $\Gamma_K = \mathrm{SL}_2(\mathfrak{O}_K)$, where $\mathfrak{O}_K$ denotes the ring of integers of K which is isomorphic to $\mathbb{Z}^n$ as additive group. The group Γ_K acts discontinuously on the product of n upper half-planes. Let $\mathfrak{a} \subset \mathfrak{O}_K$ be an ideal that is different from 0. The principal congruence subgroup of level $\mathfrak{a}$ is the kernel of the natural homomorphism $\mathrm{SL}_2(\mathfrak{O}_K) \to \mathrm{SL}_2(\mathfrak{O}_K/\mathfrak{a})$. The group $\mathrm{SL}_2(\mathfrak{O}_K/\mathfrak{a})$ is finite because $\mathfrak{O}_K/\mathfrak{a}$ is a finite ring. Hence the kernel is a normal subgroup of finite index. We denote it by

$$\Gamma_K(\mathfrak{a}) = \{M \in \Gamma_K \,:\, M \equiv \mathrm{Id} \pmod{\mathfrak{a}}\}.$$

There exists an ideal $\mathfrak{a} \subset \mathfrak{O}_K$ that is different from 0 such that $\Gamma_K(\mathfrak{a}) \subset A\Gamma_K A^{-1}$ with $A \in \mathrm{GL}_2(K)$.

Next we determine the cusp classes of Γ_K. Γ_K has the cusp ∞, and the stabilizer of ∞ consists of all transformations of the form

$$z \mapsto \varepsilon^2 z + a \quad (a \in \mathfrak{O}_K,\, \varepsilon \in \mathfrak{O}_K^\times),$$

where $\mathfrak{O}_K^\times$ is the multiplicative group of units of $\mathfrak{O}_K$. It is easy to check that the cusps of Γ_K are the elements of $K \cup \{\infty\}$. And we have to clarify when two cusps are equivalent under the action of Γ_K. For this purpose, we consider for each $a \in K$ the ideal $(a, 1)$ generated by a and 1, with $a \mapsto (a, 1)$. We extend this by sending ∞ to the principal ideal (1), $\infty \mapsto (1)$. If we denote by $C(K)$ the finite set of ideal classes of K, the above construction gives a mapping $K \cup \{\infty\} \to C(K)$, which is surjective. Furthermore, two cusps are equivalent with respect to Γ_K if and only if the corresponding ideal classes coincide. Therefore, Γ_K has only finitely many cusp classes, and their number equals the class number of K. The finiteness of the number of cusp classes gives a hint towards the following result: The space $X_{\Gamma_K} = S^n \cup K \cup \{\infty\}/\Gamma_K$ is compact. As consequences of this, we obtain that Γ_K has only a finite number of conjugacy classes of elements of finite order as well as of cusps.

9.1.2 *Hilbert modular forms*

Let $r = (r_1, \ldots, r_n)$ be a vector of rational integers. For $M \in \mathrm{SL}_2(\mathbb{R})^n$ and $z \in S^n$, we put

$$\begin{aligned} J(M, z) &= N(cz + d)^r \\ &= \prod_{j=1}^{n} (c_j z_j + d_j)^{r_j}. \end{aligned}$$

Then $J(M,z)$ is a so-called *factor of automorphy*, i.e.

$$J(MN,z) = J(M,Nz)\cdot J(N,z).$$

Let t be any lattice in $\mathbb{R}^n$. The dual lattice is defined by

$$t^\circ = \{a\in\mathbb{R}^n : \mathrm{tr}(ax)\in\mathbb{Z} \text{ for all } x\in t\}.$$

Let $V\subset\mathbb{R}^n$ be any open connected domain, and put

$$D=\{z\in\mathbb{C}^n : y\in V\}$$

Let $f: D\to\mathbb{C}$ be any holomorphic function on the *tube domain* D over V. Assume that

$$f(z+a)=f(z),\quad a\in t,$$

where $t\subset\mathbb{R}^n$ is some lattice. Then f has a unique Fourier expansion

$$f(z)=\sum_{g\in t^\circ} a_g e^{2\pi i \mathrm{tr}(gz)}.$$

The series converges absolutely and uniformly on compact subsets. If y is an arbitrary point of V, the formula

$$a_g=\mathrm{vol}(P)^{-1}\int_P f(z)e^{-2\pi i\mathrm{tr}(gz)}dx$$

holds, where $z=x+iy$, $dx=dx_1\cdots dx_n$ and $\mathrm{vol}(P)$ denotes the Euclidean volume of a fundamental parallelotope P of t. Now if t is the translation lattice of a discrete subgroup $\Gamma\subset\mathrm{SL}_2(\mathbb{R})^n$ with cusp ∞, then

$$t=\{a\in\mathbb{R}^n : z\mapsto z+a \text{ lies in } \Gamma\},$$

and let $f: S^n\to\mathbb{C}$ be a holomorphic function which is periodic with respect to the translation lattice t of Γ. We call f *regular at the cusp* ∞ if the Fourier coefficients satisfy the condition

$$a_g\neq 0\Longrightarrow g\geqq 0.$$

Also we say f *vanishes at cusp* ∞ if it is regular and $a_0=0$, namely

$$a_g\neq 0\Longrightarrow g>0.$$

We generalize this to an arbitrary cusp κ. Let $\Gamma\subset\mathrm{SL}_2(\mathbb{R})^n$ be a discrete subgroup with cusp κ and $f: S^n\to\mathbb{C}$ be a holomorphic function with the transformation law

$$f(Mz)=J(M,z)f(z),$$
$$J(M,z)=N(cz+d)^r,\quad (r\in\mathbb{Z}^n)$$

for all $M \in \Gamma$. Then we say that f *is regular at the cusp* κ (resp. *vanishes at* κ) if for some matrix

$$A \in \mathrm{SL}_2(\mathbb{R})^n, \quad A\kappa = \infty,$$

and $f_A(z) = J(A^{-1}, z)f(A^{-1}z)$, f_A is regular at ∞ (resp. vanishes at ∞).

Now we define the following *Hilbert modular forms.* Let Γ be a subgroup of Γ_K such that the quotient $(S^n)^*/\Gamma$ is compact. A Hilbert modular form of weight r, $r = (r_1, \ldots, r_n) \in \mathbb{Z}^n$ with respect to Γ is a holomorphic function $f : S^n \to \mathbb{C}$ with the properties

1) $f(Mz) = N(cz + d)^r f(z), \quad M \in \Gamma,$
2) f is regular at the cusps.

If f vanishes at all of the cusps, we call f a Hilbert cusp form. It is sufficient to verify condition 2) only for a set of representatives of all of the cusps.

9.2 A dimension formula for the space of the Hilbert cusp forms of weight 1 of two variables

9.2.1 *Introduction*

The dimension of the space of Hilbert cusp forms has been calculated in most of the cases, but not yet for the case of weight 1. In this section we shall treat some dimension formula for this remaining case under some restriction, by using the *Selberg trace formula.* Firstly we shall introduce some notation:

$$\begin{aligned}
S^2 &= \{w = (z_1, z_2) \,:\, z_j \in S \quad (j = 1, 2)\}, \\
\tilde{S} &= \{\tilde{z} = (z, \phi) \,:\, z \in S,\, \phi \in T\}, \quad T\text{: real torus}, \\
\tilde{S}^2 &= \left\{\tilde{w} = (\tilde{z}_1, \tilde{z}_2) \,:\, \tilde{z}_j \in \tilde{S}, \quad (j = 1, 2)\right\},
\end{aligned}$$

$$\begin{aligned}
G &= \mathrm{SL}_2(\mathbb{R}), \quad \tilde{G} = G \times T, \\
G^2 &= G \times G, \quad \tilde{G}^2 = \tilde{G} \times \tilde{G}.
\end{aligned}$$

The operation of $\sigma = (\sigma_1, \sigma_2) \in \tilde{G}^2$ on $\tilde{S}^2$ is represented as follows:

$$\begin{aligned}
\sigma(\tilde{w}) &= (\sigma_1\tilde{z}_1, \sigma_2\tilde{z}_2) \\
\sigma_j\tilde{z}_j &= (g_j, \alpha_j)(z_j, \phi_j) \\
&= \left(\frac{a_j z_j + b_j}{c_j z_j + d_j}, \phi_j + \arg(c_j z_j + d_j) - \alpha_j\right)
\end{aligned}$$

for $g_j = \begin{pmatrix} a_j & b_j \\ c_j & d_j \end{pmatrix}$ and $\alpha_j \in T$ with $j = 1, 2$.

The $\tilde{G}^2$-*invariant metric* on $\tilde{S}^2$ is given by

$$ds^2 = \sum_{j=1}^{2} \left(\frac{dx_j + dy_j}{y_j^2} + \left(d\phi_j - \frac{dx_j}{2y_j} \right)^2 \right),$$

and *the* $\tilde{G}^2$-*invariant measure* $d\tilde{w}$ associated to ds^2 is given by

$$d\tilde{w} = \prod_{j=1}^{2} \frac{dx_j dy_j d\phi_j}{y_j^2}, \quad (z_j = x_j + \sqrt{-1} y_j,\ j = 1, 2).$$

The *ring of* $\tilde{G}^2$-*invariant differential operators* on $\tilde{S}^2$ is generated by

$$\frac{\partial}{\partial \phi_j}$$

and

$$\tilde{\Delta}_j = y_j^2 \left(\frac{\partial^2}{\partial x_j^2} + \frac{\partial^2}{\partial y_j^2} \right) + \frac{5}{4} \frac{\partial^2}{\partial \phi_j^2} + y_j \frac{\partial}{\partial \phi_j} \frac{\partial}{\partial x_j} \quad (j = 1, 2).$$

Let K be a real quadratic field and $\mathfrak{O}_K$ be the ring of integers of K. We put

$$\mathrm{SL}_2(\mathfrak{O}_K) = \left\{ \begin{pmatrix} a & b \\ c & d \end{pmatrix} : ad - bc = 1,\ a, b, c, d \in \mathfrak{O}_K \right\}.$$

Then the Hilbert modular group associated to K is

$$\Gamma_K = \left\{ \left(\begin{pmatrix} a_1 & b_1 \\ c_1 & d_1 \end{pmatrix}, \begin{pmatrix} a_2 & b_2 \\ c_2 & d_2 \end{pmatrix} \right) : \begin{pmatrix} a_1 & b_1 \\ c_1 & d_1 \end{pmatrix} \in \mathrm{SL}_2(\mathfrak{O}_K) \right\},$$

where a_2, b_2, c_2, d_2 denote the conjugate of a_1, b_1, c_1, d_1 respectively. It is well known that its number of cusps equals the class number $h(K)$ of K. Let Γ be a congruence subgroup of Γ_K, and for $\gamma = (\gamma_1, \gamma_2) = \left(\begin{pmatrix} a_1 & b_1 \\ c_1 & d_1 \end{pmatrix}, \begin{pmatrix} a_2 & b_2 \\ c_2 & d_2 \end{pmatrix} \right) \in \Gamma$ and $w \in S^2$, we put

$$J(\gamma, w) = (c_1 z_1 + d_1)(c_2 z_2 + d_2).$$

We say that a function $f(w)$ defined on S^2 is a *Hilbert cusp form of weight* 1 for Γ if

1) $f(z)$ is holomorphic on S^2,

2) $f(\gamma w) = J(\gamma, w) f(w)$ for all $\gamma \in \Gamma$,

3) At every parabolic point κ of Γ, a constant term in the Fourier expansion of $f(w)$ at κ vanishes.

We denote by $S_1(\Gamma)$ the space of Hilbert cusp forms of weight 1 for Γ and put

$$d_1 = \dim S_1(\Gamma).$$

In this section, we shall calculate the dimension d_1 for some Γ.

Here we give some historical remarks. For the case of n complex variables and the weight $r = (r_1, \ldots, r_n)$, the dimension formulas have been calculated in the following:

- all $r_j > 2$ and even n case by H. Shimizu (1963, [94]) and R. Busam (1970);
- all $r_j = 2$ case by K.-B. Merz (1971) and E. Freitag (1972);
- all $r_j = 2$ and $n = 2$ case by F. Hirzeburch (1973, by algebraic geometric method);
- all $r_j \geqq 2$ case by H. Ishikawa (1974, [52]).

We consider the case of all $r_j = 1$ and $n = 2$.

9.2.2 *Fundamental lemma*

We denote by $\mathfrak{M}_\Gamma((k_1, \lambda_1), (k_2, \lambda_2))$ the set of all functions $F(\tilde{w})$ satisfying the following conditions:

i) $F(\tilde{w}) \in L^2(\Gamma \setminus \tilde{S}^2)$,
ii) For $j = 1, 2$

$$\tilde{\Delta}_j F(\tilde{w}) = \lambda_j F(\tilde{w}), \tag{9.1}$$

$$\frac{\partial}{\partial \phi_j} F(\tilde{w}) = -\sqrt{-1} k_j F(\tilde{w}). \tag{9.2}$$

Then we have the following fundamental lemma.

Lemma 9.1. *The notation being as above, we have*

$$\mathfrak{M}_\Gamma\left(\left(1, -\frac{3}{2}\right), \left(1, -\frac{3}{2}\right)\right)$$

$$= \left\{e^{-\sqrt{-1}(\phi_1+\phi_2)} y_1^{\frac{1}{2}} y_2^{\frac{1}{2}} f(z_1, z_2) \,:\, f(z_1, z_2) \in S_1(\Gamma)\right\},$$

and hence

$$d_1 = \dim \mathfrak{M}_\Gamma \left(\left(1, -\frac{3}{2}\right), \left(1, -\frac{3}{2}\right) \right).$$

Proof. For each $f(z_1, z_2) \in S_1(\Gamma)$, we put

$$F(\tilde{w}) = e^{-\sqrt{-1}(\phi_1+\phi_2)} y_1^{\frac{1}{2}} y_2^{\frac{1}{2}} f(z_1, z_2). \tag{9.3}$$

Then it is easy to check that the function $F(\tilde{w})$ belongs to $\mathfrak{M}_\Gamma\left(\left(1, -\frac{3}{2}\right), \left(1, -\frac{3}{2}\right)\right)$.

We now prove that conversely any element in $\mathfrak{M}_\Gamma\left(\left(1, -\frac{3}{2}\right), \left(1, -\frac{3}{2}\right)\right)$ must be of the form (9.3) with $f(z_1, z_2) \in S_1(\Gamma)$. Put

$$f(z_1, z_2) = e^{\sqrt{-1}(\phi_1+\phi_2)} y_1^{-\frac{1}{2}} y_2^{-\frac{1}{2}} F(\tilde{w}).$$

In the following we shall prove that $f(z_1, z_2)$ belongs to $S_1(\Gamma)$. Firstly the Γ-invariance of $F(\tilde{w})$ is equivalent to

$$f(\gamma w) = J(\gamma, w) f(w)$$

for all $\gamma \in \Gamma$. Therefore it is sufficient for the proof of the latter half of Lemma 9.1 to show that $f(w)$ satisfies the conditions 1) and 3) in Subsection 9.2.1.

Let $\mathbb{A}$ be the adele ring of the real quadratic field K. Let G_K be $\mathrm{GL}_2(K)$ which is viewed as an algebraic group over K and $G_\mathbb{A}$ the adelization of G_K. The Hilbert modular forms may be viewed as automorphic forms on $G_\mathbb{A}$. Firstly we put

$$G_\mathbb{A} = G_{\mathbb{A}_f} \times G_{\mathbb{A}_\infty},$$

where $G_{\mathbb{A}_f}$ (resp. $G_{\mathbb{A}_\infty}$) denotes the finite (resp. infinite) component of $G_\mathbb{A}$, and we put

$$K_f = \text{ open compact subgroup of } G_{\mathbb{A}_f},$$
$$G_\infty^+ = \mathrm{GL}_2(\mathbb{R})^+ \times \mathrm{GL}_2(\mathbb{R})^+,$$

and

$$\Gamma = G_K \cap \left(K_f \times G_\infty^+\right).$$

Then Γ is a discrete subgroup of G_∞^+, and we have

$$G_\mathbb{A} = \bigcup_{k=1}^{h} G_K x_k \left(K_f \times G_\infty^+\right) \quad \text{(disjoint)}.$$

Now we put

$$\Gamma_k = G_K \cap \left(x_k K_f \times G_\infty^+\right),$$
$$M_k = \mathfrak{M}_{\Gamma_k} \left(\left(1 - \frac{3}{2}\right), \left(1 - \frac{3}{2}\right) \right)$$

with $1 \leqq k \leqq h$. Let M be the space of all φ on $G_\mathbb{A}$ satisfying the following conditions (1)–(3):

(1) $\varphi(\alpha x k_f k_\infty t) = e^{\sqrt{-1}(\phi_1+\phi_2)}\varphi(x)$,
where $\alpha \in G_K$, $k_f \in K_f$, $k_\infty = (k(\phi_1), k(\phi_2)) \in K_\infty = \mathrm{SO}(2) \times \mathrm{SO}(2)$, and

$$t \in Z_\infty^+ = \left\{ \left(\begin{pmatrix} t_1 & 0 \\ 0 & t_1 \end{pmatrix}, \begin{pmatrix} t_2 & 0 \\ 0 & t_2 \end{pmatrix} \right) : t_j > 0 \quad (j = 1, 2) \right\};$$

(2) $\tilde{\Delta}_j \varphi = \lambda_j \varphi$ as function of x ($\lambda_1 = \lambda_2 = -\frac{3}{2}$);

(3) $\displaystyle\int_{Z_\infty^+ G_K \backslash G_{\mathbb{A}}} |\varphi(g)|^2 dg < \infty$.

For $x \in G_{\mathbb{A}}$, we put $x = \alpha\, x_k\, k_f\, g$ ($\alpha \in G_K$, $k_f \in K_f$, $g \in G_\infty^+$) and

$$\varphi(x) = F_j(g\tilde{w}_0) \quad (\tilde{w}_0 = ((\sqrt{-1}, 0), (\sqrt{-1}, 0))).$$

Then, $(F_1, \ldots, F_h) \mapsto \varphi$ gives an isomorphism of $M_1 \times \cdots \times M_h$ onto M. Let Z be the center of GL_2, and ω a character of $Z_{\mathbb{A}}/Z_K(Z_{\mathbb{A}} \cap K_f)Z_\infty^+$. We put

$$L^2(\omega, \mathrm{GL}_2) = \left\{ \varphi : G_{\mathbb{A}} \to \mathbb{C} \;\middle|\; \begin{array}{l} \varphi(a\alpha x) = \omega(a)\varphi(x) \text{ for all } \alpha \in G_K, \\ \int_{Z_{\mathbb{A}} G_K \backslash G_{\mathbb{A}}} |\varphi(x)|^2 dx < \infty \end{array} \right\}.$$

Then the space $\bigoplus_\omega L^2(\omega, \mathrm{GL}_2)$ contains the space M. By [26], we have now that $L^2(\omega, \mathrm{GL}_2)$ decomposes as the direct sum

$$L^2(\omega, \mathrm{GL}_2) = L^2_{\mathrm{cusp}}(\omega, \mathrm{GL}_2) \oplus L^2_{\mathrm{sp}}(\omega, \mathrm{GL}_2) \oplus L^2_{\mathrm{cont}}(\omega, \mathrm{GL}_2),$$

where

$$L^2_{\mathrm{cusp}}(\omega, \mathrm{GL}_2) = \left\{ \varphi \in L^2(\omega, \mathrm{GL}_2) : \int_{N_{\mathbb{A}}/N_K} \varphi(ng)dn = 0 \text{ for all } g \in G_{\mathbb{A}} \right\}$$

with $N = \left\{ \begin{pmatrix} 1 & b \\ 0 & 1 \end{pmatrix} \right\}$, L^2_{sp} is the space spanned by χ $(\det g)$ with a Grössen -character of K such that $\chi^2 = \omega$. Then it is known in general the following (A) and (B) ([26]).

(A) The space $L^2_{\mathrm{cont}}(\omega, \mathrm{GL}_2)$ which is the continuous part of $L^2(\omega, \mathrm{GL}_2)$, does not contain the eigenfunctions of the Laplacian $\tilde{\Delta}_j - \dfrac{5}{4}\dfrac{\partial^2}{\partial \phi_j^2}$ $(j = 1, 2)$;

(B) If $F \in M$, then the function $f(z_1, z_2)$ is holomorphic.

By (A) and (B), we have completed the proof of the latter half of Lemma 9.1. □

9.2.3 *Modified trace formula*

We put $\boldsymbol{\lambda}_j = (k_j, \lambda_j)$ $(j = 1, 2)$. For every invariant integral operator with a kernel $k(\tilde{w}; \tilde{v})$ on $\mathfrak{M}_\Gamma(\boldsymbol{\lambda}_1, \boldsymbol{\lambda}_2)$, we have

$$\int_{\tilde{S}^2} k(\tilde{w}; \tilde{v}) f(\tilde{v}) d\tilde{v} = \tilde{h}(\boldsymbol{\lambda}_1, \boldsymbol{\lambda}_2) f(\tilde{w})$$

for $f \in \mathfrak{M}_\Gamma(\boldsymbol{\lambda}_1, \boldsymbol{\lambda}_2)$. Put

$$\tilde{K}(\tilde{w}; \tilde{v}) = \sum_{\gamma \in \Gamma} k(\tilde{w}; \gamma\tilde{v}),$$

then

$$\int_{\Gamma \backslash \tilde{S}^2} \tilde{K}(\tilde{w}; \tilde{v}) f(\tilde{v}) d\tilde{v} = \tilde{h}(\boldsymbol{\lambda}_1, \boldsymbol{\lambda}_2) f(\tilde{w}).$$

Denote by $\Gamma(\gamma)$ the centralizer of γ in Γ and put $\tilde{\mathscr{F}}_{\Gamma(\gamma)} = \Gamma(\gamma) \backslash \tilde{S}^2$. Then it is easy to see that formally,

$$\begin{aligned}\int_{\Gamma \backslash \tilde{S}^2} \tilde{K}(\tilde{w}; \tilde{v}) d\tilde{w} &= \sum_{\{\gamma\}} \sum_{\sigma \in \Gamma / \Gamma(\gamma)} \int_{\Gamma \backslash \tilde{S}^2} k(\tilde{w}; \sigma^{-1}\gamma\sigma\tilde{w}) d\tilde{w} \\ &= \sum_{\{\gamma\}} \int_{\tilde{\mathscr{F}}_{\Gamma(\gamma)}} k(\tilde{w}; \gamma\tilde{v}) d\tilde{w},\end{aligned}$$

where the sum $\{\gamma\}$ is taken over the distinct conjugacy classes of Γ.

Since our Γ is of finite covolume type, its spectrum has a continuous part, and the *continuous spectrum* can be described by a family of Eisenstein series defined in Subsection 9.2.4 below. Using the Eisenstein series, we shall construct in the subsequent section a new kernel H_δ related with some $k(\tilde{w}; \tilde{v})$. Then $\tilde{K} - H_\delta$ is a *Hilbert-Schmidt kernel* (cf. Subsection 9.2.4). Therefore we have the following modified Selberg trace formula

$$\sum_{j=1}^{\infty} \tilde{h}(\boldsymbol{\lambda}_1^{(j)}, \boldsymbol{\lambda}_2^{(j)}) = \int_{\Gamma \backslash \tilde{S}^2} \{K(\tilde{w}; \tilde{w}) - H_\delta\} d\tilde{w}. \qquad (*)$$

Now we consider the following invariant integral operator defined by

$$\omega_\delta(\tilde{w}; \tilde{v}) = \prod_{j=1}^{2} \left\{ \left| \frac{(y_j y_j')^{\frac{1}{2}}}{(z_j - \bar{z}_j')/2\sqrt{-1}} \right|^\delta \frac{(y_j y_j')^{\frac{1}{2}}}{(z_j - \bar{z}_j')/2\sqrt{-1}} e^{-\sqrt{-1}(\phi_j - \phi_j')} \right\},$$

where $\tilde{v} = (\tilde{z}_1', \tilde{z}_2')$, $\operatorname{Re} \delta > 1$. The integral operator ω_δ vanishes on $\mathfrak{M}_\Gamma(\boldsymbol{\lambda}_1, \boldsymbol{\lambda}_2)$ for all $(\boldsymbol{\lambda}_1, \boldsymbol{\lambda}_2)$ except $k_1 = k_2 = 1$. We denote by

$$\mu_{\alpha\beta} = \Big((1, \lambda^{(\alpha)}), (1, \lambda^{(\beta)}) \Big), \quad (\alpha, \beta \geqq 2)$$

and

$$\mu_{11} = \left(\left(1, -\frac{3}{2}\right), \left(1, -\frac{3}{2}\right)\right), \quad \lambda^{(1)} = -\frac{3}{2}$$

the discrete part of spectra, and we put

$$d_{\alpha\beta} = \dim \mathfrak{M}_\Gamma(\mu_{\alpha\beta}).$$

Then the left-hand side of $(*)$ implies that

$$\sum_{j=1}^{\infty} \tilde{h}(\boldsymbol{\lambda}_1^{(j)}, \boldsymbol{\lambda}_2^{(j)}) = \sum_{\alpha,\beta=1}^{\infty} d_{\alpha\beta}\Lambda_{\alpha\beta}$$

where $\Lambda_{\alpha\beta}$ denotes the eigenvalue of ω_δ in $\mathfrak{M}_\Gamma(\mu_{\alpha\beta})$. For the eigenvalue $\Lambda_{\alpha\beta}$ we have

$$\Lambda_{\alpha\beta} = \left\{2^{2+\delta}\pi \frac{\Gamma(\frac{1}{2})\Gamma(\frac{\delta+1}{2})}{\Gamma(\delta)\Gamma(1+\frac{\delta}{2})}\right\}^2$$

$$\times \Gamma\left(\frac{\delta}{2} + \sqrt{-1}u_\alpha\right)\Gamma\left(\frac{\delta}{2} - \sqrt{-1}u_\alpha\right)\Gamma\left(\frac{\delta}{2} + \sqrt{-1}u_\beta\right)\Gamma\left(\frac{\delta}{2} - \sqrt{-1}u_\beta\right),$$

where $\lambda^{(\ell)} = r_\ell(r_\ell - 1) - \frac{5}{4}$ and $r_\ell = \frac{1}{2} + \sqrt{-1}u_\ell$ with $\ell = \alpha, \beta$ ([14], [15]). In general, it is known that the series $\sum_{\alpha,\beta=1}^{\infty} d_{\alpha,\beta}\Lambda_{\alpha\beta}$ is absolutely convergent for $\operatorname{Re}\delta > 1$, and by Stirling formula, we see that the above series is absolutely and uniformly convergent for all bounded δ except for $\delta = \pm(2r_\alpha - 1), \pm(2r_\beta - 1)$. Note that $\delta = 0$ if and only if $(\lambda^{(\alpha)}, \lambda^{(\beta)}) = (\lambda^{(1)}, \lambda^{(1)})$. In the following, we discuss the contributions from the compact part to the modified trace formula $(*)$.

1) $\gamma =$ identity.

$$\int_{\tilde{\mathscr{F}}_{\Gamma(\gamma)}} d\tilde{\omega} = \int_{\tilde{\mathscr{F}}_\Gamma} d\tilde{\omega} < \infty \quad (\tilde{\mathscr{F}}_\Gamma = \Gamma \setminus \tilde{S}^2).$$

2) $\gamma =$ elliptic.
We put $\gamma = (\gamma_1, \gamma_2)$, and let ζ_j, $\bar{\zeta}_j$ be the eigenvalues of γ_j. Consider a linear transformation that maps S^2 into the product of the 2 unit circles, and a fixed point of γ to the origin. Then

$$\frac{\gamma_j z_j - \rho_j}{\gamma_j z_j - \bar{\rho}_j} = \frac{\zeta_j}{\bar{\zeta}_j} \cdot \frac{z_j - \rho_j}{z_j - \bar{\rho}_j} \quad (j = 1, 2),$$

where $\rho = (\rho_1, \rho_2) \in S^2$ is the fixed point of γ. By a simple calculation which is similar to that of elliptic modular case, we have the following contribution from this part:

$$J(\gamma) = \sum_{\{\gamma\}} \frac{(8\pi^2)^2}{[\Gamma(\gamma) : Z(\Gamma)]} \frac{1}{\delta^2} \prod_{j=1}^{2} \frac{\bar{\zeta}_j}{1-\bar{\zeta}_j^2} \cdot F\left(1, 1+\frac{\delta}{2}, 1+\delta; \frac{\bar{\zeta}_j}{1-\bar{\zeta}_j^2}\right),$$

and therefore

$$\lim_{\delta \to +0} \delta^2 J(\gamma) = (8\pi^2)^2 \sum_{\{\gamma\}} \frac{1}{[\Gamma(\gamma) : Z(\Gamma)]} \prod_{j=1}^{2} \frac{\bar{\zeta}_j}{1-\bar{\zeta}_j^2},$$

where $Z(\Gamma)$ denotes the center of Γ, the sum $\{\gamma\}$ is taken over the totally elliptic conjugacy classes in Γ and $F(*)$ denotes a *hypergeometric function of Gauss.*

3) $\gamma =$ totally hyperbolic.
The contribution of this part is essential in the case of weight 1. We put $\gamma = (\gamma_1, \gamma_2)$. Then there exists some g_j in G such that

$$g_j^{-1} \gamma_j g_j = \begin{pmatrix} \lambda_{0j} & 0 \\ 0 & \lambda_{0j}^{-1} \end{pmatrix}, \quad |\lambda_{0j}| > 1 \quad (j = 1, 2).$$

Now, let $\{\gamma_1, \gamma_2\}$ be a system of generators of $\Gamma(\gamma)$ and $\lambda_0^{(\ell)} = (\lambda_{01}^{(\ell)}, \lambda_{02}^{(\ell)})$ $(|\lambda_{0j}^{(\ell)}| > 1,\ \ell = 1, 2)$ denotes an eigenvalue of γ_ℓ respectively. Writing $z_j = \rho_j \exp(\sqrt{-1}\theta_j)$ and $\log \rho_j = u_1 \log \lambda_{0j}^{(1)} + u_2 \log \lambda_{0j}^{(2)}$ with $u_\ell \in \mathbb{R}$, the set of $z = (z_1, z_2)$ such that

$$0 < u_\ell < 1 \quad (\ell = 1, 2), \quad 0 < \theta_j < \pi \quad (j = 1, 2)$$

forms a fundamental domain of $\Gamma(\gamma)$ in S^2. The contribution from this part is as follows ([15]):

$$J(\gamma) = 2^{6+2\delta}\pi^3 \left\{ \frac{\Gamma(\frac{\delta+1}{2})}{\Gamma(\frac{\delta+2}{2})} \right\}^2 \sum_{\{\gamma\} \in P} \prod_{j=1}^{2} \frac{\operatorname{sgn}(\lambda_j) \cdot \mu(\gamma)}{|\lambda_j + \lambda_j^{-1}|^\delta |\lambda_j - \lambda_j^{-1}|},$$

where P denotes a complete system of totally hyperbolic conjugacy classes in Γ such that none of its fixed points is a parabolic point of Γ, $\lambda = (\lambda_1, \lambda_2)$ is an eigenvalue of γ and $\mu(\gamma) = \det\left(\log|\lambda_{0j}^{(\ell)}|\right)_{j,\ell=1,2}$.

4) $\gamma =$ mixed.
We give the contributions from 3) and 4) by unified form. Let $\gamma \in \Gamma$ be an element with m hyperbolic and $n-m$ elliptic ($m = 1$ or 2, $n = 2$).

Then $\Gamma(\gamma)$ is free abelian group of rank m, say $\Gamma(\gamma) = \langle \gamma_1, \ldots, \gamma_m \rangle$, and we have the following contribution from 3) and 4):

$$J_\delta(\gamma) = \sum_{m=1}^{n} \left\{ 2^{3+\delta}\pi \frac{\Gamma(\frac{1}{2})\Gamma(\frac{1+\delta}{2})}{\Gamma(1+\frac{\delta}{2})} \right\}^m \left(\frac{1}{\delta}\right)^{n-m}$$

$$\times \sum_{\{\gamma\} \in P_m} \mathrm{vol}(\mathscr{F}_{\Gamma(\gamma)}) \prod_{1 \leqq j \leqq m} \frac{\mathrm{sgn}(\lambda_j)}{|\lambda_j + \lambda_j^{-1}|^\delta |\lambda_j - \lambda_j^{-1}|}$$

$$\times \prod_{m+1 \leqq j \leqq n} \bar{\zeta}_j F_1\left(1, 1+\frac{\delta}{2}, \frac{\delta}{2}, 1+\delta; \bar{\zeta}_j^2, \zeta_j^2\right),$$

where $F_1(*)$ denotes the *hypergeometric function of two variables*, and P_m denotes a complete system of conjugacy classes of m hyperbolic and $n-m$ elliptic elements in Γ. We put

$$Z(\delta) = (16\pi^2)^{-2} J_\delta(\gamma).$$

Then, by the trace formula $(*)$, the function $Z(\delta)$ will be extended to a meromorphic function on the half-plane $\mathrm{Re}\,\delta > 0$.

9.2.4 *Eisenstein series attached to* ∞

For the sake of simplicity, we shall assume that $\mathscr{F}_\Gamma$ has only one cusp, i.e. the cusp is at $\infty = (\infty, \infty)$. Let $s \in \mathbb{C}$, $m \in \mathbb{Z}$ and Γ_∞ be the stabilizer of ∞ in Γ. Then the *Eisenstein series attached to* ∞ is defined by

$$E(\tilde{w}; s, m) = \sum_{\substack{M \in \Gamma_\infty \backslash \Gamma \\ M = (M_1, M_2)}} \{y(M_1 z_1) y(M_2 z_2)\}^s$$
$$\times e^{-\sqrt{-1}\{\phi_1 + \phi_2 + \arg(c_1 z_1 + d_1)(c_2 z_2 + d_2)\}}$$
$$\times \{y(M_1 z_1)\}^{\frac{\pi\sqrt{-1}m}{2\log\varepsilon}} \{y(M_2 z_2)\}^{-\frac{\pi\sqrt{-1}m}{2\log\varepsilon}},$$

where, ε denotes the fundamental unit (> 1) of K. $y(z) = \mathrm{Im}\, z$, and $M_j = \begin{pmatrix} a_j & b_j \\ c_j & d_j \end{pmatrix}$ $(j = 1, 2)$. The series $E(\tilde{w}; s, m)$ converges absolutely for $\mathrm{Re}\, s > 1$, and has the following properties:

$$E(\gamma\tilde{w}; s, m) = E(\tilde{w}; s, m)$$

for $\gamma \in \Gamma$, and

$$\frac{\partial}{\partial \phi_j} E = -\sqrt{-1} E, \quad \tilde{\Delta}_j E = \lambda_j E \quad (j = 1, 2),$$

where

$$\lambda_1 = \left(s + \frac{\pi}{2}\frac{\sqrt{-1}m}{\log \varepsilon}\right)\left(s + \frac{\pi}{2}\frac{\sqrt{-1}m}{\log \varepsilon} - 1\right) - \frac{5}{4},$$

$$\lambda_2 = \left(s - \frac{\pi}{2}\frac{\sqrt{-1}m}{\log \varepsilon}\right)\left(s - \frac{\pi}{2}\frac{\sqrt{-1}m}{\log \varepsilon} - 1\right) - \frac{5}{4}.$$

We set

$$E(w; s, m) = e^{\sqrt{-1}(\phi_1+\phi_2)} E(\tilde{w}; s, m).$$

Then the series $E(w; s, m)$ is invariant under the action of the lattice $\mathfrak{O}_K$, and therefore has a Fourier expansion of the form

$$E(w; s, m) = \sum_{\substack{\ell \in \mathfrak{O}_K^* \\ \ell = (\ell_1, \ell_2)}} a_\ell(y; s, m) e^{2\sqrt{-1}\pi\langle \ell, x\rangle},$$

where $\langle \ell, x\rangle = \ell_1 x_1 + \ell_2 x_2$, and $\mathfrak{O}_K^*$ is the dual lattice of $\mathfrak{O}_K$, i.e. $\mathfrak{O}_K^* = \{\alpha \in K : \mathrm{tr}(\alpha\mathfrak{O}_K) \subset \mathbb{Z}\}$. The constant term $a_0(y; s, m)$ is given by

$$y_1^{s+\frac{\pi}{2}\frac{\sqrt{-1}m}{\log\varepsilon}} y_2^{s-\frac{\pi}{2}\frac{\sqrt{-1}m}{\log\varepsilon}} - \varphi(s, m) y_1^{1-s-\frac{\pi}{2}\frac{\sqrt{-1}m}{\log\varepsilon}} y_2^{1-s+\frac{\pi}{2}\frac{\sqrt{-1}m}{\log\varepsilon}},$$

where D is the discriminant of K,

$$\varphi(s, m) = \frac{\pi}{\sqrt{D}} \frac{\Gamma\left(s + \frac{\pi}{2}\frac{\sqrt{-1}m}{\log\varepsilon}\right)\Gamma\left(s - \frac{\pi}{2}\frac{\sqrt{-1}m}{\log\varepsilon}\right)}{\Gamma\left(s + \frac{\pi}{2}\frac{\sqrt{-1}m}{\log\varepsilon} + \frac{1}{2}\right)\Gamma\left(s - \frac{\pi}{2}\frac{\sqrt{-1}m}{\log\varepsilon} + \frac{1}{2}\right)} \frac{L(2s-1, -m)}{L(2s, -m)},$$

$$\xi_{-m}(c) = \left|\frac{c}{c'}\right|^{-\frac{\pi\sqrt{-1}m}{\log\varepsilon}}$$ is the Grössencharacter of K and,

$$L(s, -m) = \sum_{\substack{(c):\ \text{ideal in } \mathfrak{O}_K \\ c \neq 0}} \mathrm{sgn}(cc')\xi_{-m}(c)|N(c)|^{-s}$$

with conjugate c' of c. This is obtained by performing calculations similar to those described in [20] and [112].

Now, by using the analytic continuation of the Eisenstein series $E(\tilde{w}; s, m)$ as a function of s for $s = \frac{1}{2} + ir$ ([20]), we put

$$H_\delta(\tilde{w}; \tilde{v}) = \frac{1}{16\pi\sqrt{D}\log\varepsilon} \sum_{m=-\infty}^{\infty} \int_{-\infty}^{\infty} \tilde{h}_\delta\left(r + \frac{\pi m}{2\log\varepsilon}, r - \frac{\pi m}{2\log\varepsilon}\right)$$

$$\times E\left(\tilde{w}, \frac{1}{2} + \sqrt{-1}r, m\right) E\left(\tilde{v}, \frac{1}{2} - \sqrt{-1}r, -m\right) dr,$$

where

$$\tilde{h}_\delta(r_1, r_2) = \left\{ 2^{2+\delta}\pi \frac{\Gamma(\frac{1}{2})\Gamma(\frac{\delta+1}{2})}{\Gamma(\delta)\Gamma(1+\frac{\delta}{2})} \right\}^2 \Gamma\left(\frac{\delta}{2} + \sqrt{-1}r_1\right)$$

$$\times \Gamma\left(\frac{\delta}{2} - \sqrt{-1}r_1\right) \Gamma\left(\frac{\delta}{2} + \sqrt{-1}r_2\right) \Gamma\left(\frac{\delta}{2} - \sqrt{-1}r_2\right).$$

Then the integral operator $\tilde{K} - H_\delta$ is now completely continuous on $L^2(\Gamma \setminus \tilde{S}^2)$ and has all of the discrete spectra of $\tilde{K}$ ([112]).

9.2.5 *The trace at the cusp*

5) $\Gamma =$ totally parabolic. Let $J(\infty)$ be the cusp contribution from $\tilde{K}$. Then, we have the following which is obtained in a similar way as in the elliptic modular case ([48]):

$$\lim_{\delta \to +0} \lim_{Y \to \infty} \delta^2 \{ J(\infty) - 2 \log \varepsilon \cdot g_\delta(0,0) \log Y \} = 0,$$

where

$$g_\delta(u_1, u_2) = \frac{1}{(2\pi)^2} \int_{-\infty}^{\infty} \int_{-\infty}^{\infty} \tilde{h}_\delta(r_1, r_2) e^{-\sqrt{-1}(r_1 u_1 + r_2 u_2)} dr_1 dr_2.$$

Therefore the contribution from any parabolic classes to d_1 vanishes.

6) $\gamma =$ hyperbolic-parabolic.
γ is conjugate in Γ to $\gamma_{m,\alpha}$:

$$\gamma \underset{\Gamma}{\sim} \gamma_{m,\alpha} = \begin{pmatrix} \varepsilon^m & \alpha \\ 0 & \varepsilon^{-m} \end{pmatrix}, \quad (m \in \mathbb{Z},\ m \neq 0,\ \alpha \in \mathfrak{O}_K).$$

The common fixed points of every element in $\Gamma(\gamma)$ are given by

$$\left\{ \infty, \frac{\alpha}{\varepsilon^{-m} - \varepsilon^m} \right\},$$

and there exists a $\tau \in \Gamma$ such that

$$\tau : \frac{\alpha}{\varepsilon^{-m} - \varepsilon^m} \longmapsto \infty.$$

We denote by $\mathscr{F}_{\Gamma(\gamma_{m,\alpha})}$ a fundamental domain $\Gamma(\gamma_{m,\alpha})$. Take Y sufficiently large and we put

$$\mathscr{F}^*_{\Gamma(\gamma_{m,\alpha})} = \left\{ w = (z_1, z_2) \in \mathscr{F}_{\Gamma(\gamma_{m,\alpha})} : y_1 y_2 \leqq Y,\ y_1' y_2' \leqq Y \right\},$$

where $\tau w = w' = (z_1', z_2')$. Moreover we put

$$J^*(\gamma) = 4\pi^2 \sum_{\{\gamma\}} \int_{\mathscr{F}^*_{\Gamma(\gamma)}} \omega_\delta(w; \gamma w) dw.$$

Then the contribution from this part is

$$\lim_{\delta \to +0} \lim_{Y \to 0} \delta^2 \left\{ J^*(\gamma) - 4 \log \varepsilon \sum_{m=1}^{\infty} \frac{g_\delta(2m \log \varepsilon,\, 2m \log \varepsilon')}{|(\varepsilon^m - \varepsilon^{-m})(\varepsilon'^m - \varepsilon'^{-m})|} \log Y \right\}$$

$$= 0.$$

7) $\operatorname{tr} H_\delta$.

By using the Maass-Selberg relation, the following contribution may be obtained in a way similar to the proof in the elliptic modular case ([20], [48]):

$$-\frac{1}{4}\tilde{h}_\delta(0,0)\varphi\left(\frac{1}{2}, 0\right).$$

Therefore

$$\lim_{\delta \to +0} \delta^2 \left\{ -\frac{1}{4}\tilde{h}_\delta(0,0)\varphi\left(\frac{1}{2}, 0\right) \right\} = -(8\pi^2)^2 \varphi\left(\frac{1}{2}, 0\right).$$

By the functional equation of $L(s, 0)$, we have

$$\varphi\left(\frac{1}{2}, 0\right) = \frac{\pi}{\sqrt{D}} \frac{\Gamma(\frac{1}{2})^2 L(0,0)}{\Gamma(1)^2 L(1,0)} = 1.$$

Finally, multiply the both side of $(*)$ by δ^2 and let δ tends to zero, then the limit express the following dimension formula:

Theorem 9.1 (Ishikawa-Hiramatsu). *The notation and the assumptions being as above, we have*

$$d_1 = \frac{1}{4} \sum_{\{\gamma\}} \frac{1}{[\Gamma(\gamma) : Z(\Gamma)]} \prod_{j=1}^{2} \frac{\bar{\zeta}_j}{1 - \bar{\zeta}_j^2} + \lim_{\delta \to +0} \delta^2 Z(\delta) - \frac{1}{4}.$$

Appendix

Some dimension formula and traces of Hecke operators for cusp forms of weight 1

(Göttingen talk, 1989. By Toyokazu Hiramatsu)

§ 1. Introduction

$$\Gamma\text{: a fuchsian group of the first kind,}$$
$$\chi\text{: a character of } \Gamma \text{ and } k \in \mathbb{Z}^+;$$

If k is odd and $\Gamma \ni -I$ ($I = \begin{pmatrix} 1 & 0 \\ 0 & 1 \end{pmatrix}$), then we assume that $\chi(-I) = -1$, i.e. χ is odd. Let $S_1(\Gamma)$ be the space of cusp forms of weight 1 of a complex variable on Γ with χ. If weight > 1, then, by making use of the Riemann-Roch theorem, we have the finite closed dimension formula for such a space.

Now we propose a problem:

$$d_1 = \dim S_1(\Gamma),\ d_1?$$

The Riemann-Roch theorem is not effective to compute the number d_1. First, I will mention some authors who have previously proposed this problem:

1) E. Hecke: Zur Theorie der elliptischen Modulfunktionen, Math. Annalen, **97** (1926).

   In this paper, the author wrote
   'Ob damit das volle Sytem von elliptischen Modulformen (−1)-ter Dimension gewonnen ist, ist für beliebige Stufenzahl noch immer eine offene Frage, ...'

2) Eichler: For $k = 1$ the theorem of Riemann-Roch becomes a tautology, and we have only very limited knowledge on d_1.

3) Private letter to the present author from Serre (1975):

   'I doubt that there is a formula in the usual sense. It is unlikely that a formula can be given by a closed formula.'

4) D. A. Hejhal: The Selberg Trace Formula for PSL(2, $\mathbb{R}$), I, Lecture Notes in Math., **548** (1976), p.434.
   'It is impossible to calculate d_1 using only the basic algebraic properties of Γ.'

5) S. Lang: Introduction to Modular Forms (1976), p.34

   'It is a major problem to determine the dimension in case $k = 1$. This ties up with the theory of representations and the existence of Galois extensions of the rationals (cf. Deligne-Serre).'

6) J. Tate: The general reciprocity law, the Hilbert Problem 9, Proc. Symposia Pure Math., **28** (1976), 311-322.

   'While it is relatively easy to construct modular forms of weight $k > 1$, and the Riemann-Roch theorem tells us exactly how many of them there are at each level, it is not so easy to exhibit forms of weight 1, and the Riemann-Roch formula fails to predict how many of them there are at a given level.'

Now, in the following, we give a formula of d_1 by using the Selberg trace formula. The Selberg trace formula is an indispensable tool in obtaining a formula for d_1. The essential part of our formula of d_1 consists of the contribution from the hyperbolic part which is the residue at the origin of a Selberg type zeta-function. Therefore, our problem gives an example not making of the Selberg principle.

§ 2. Results

Many mathematicians have investigated d_1. For the sake of simplicity I will discuss only my own contributions in the following.

We put

$$\zeta_{\Gamma,\chi}(s) = \sum_{\{P\}_\Gamma} \frac{(\mathrm{sgn}P)^\ell\,\chi(P)\,\log N(P_0)}{N(P)^{\frac{1}{2}} - N(P)^{-\frac{1}{2}}} \left(2\cosh\frac{\log N(P)}{2}\right)^{-2s}, \quad \mathrm{Re}\,s > \frac{1}{2},$$

where the sum over $\{P\}_\Gamma$ is taken over the distinct hyperbolic conjugacy classes of Γ, P_0 denotes the primitive hyperbolic element associated with P, i.e. $P = P_0^m$ $(m \in \mathbb{Z}^+)$, $\ell = 0$ or 1 according to $-I \in \Gamma$ or not, and if $A^{-1}PA = \begin{pmatrix} \lambda & 0 \\ 0 & \lambda^{-1} \end{pmatrix}$, $|\lambda| > 1$, then $N(P) = \lambda^2$. The Selberg type zeta-function $\zeta_{\Gamma,\chi}(s)$ is continued analytically to a meromorphic function on the whole s-plane having a simple pole at $s = 0$.

(1) The case 1: $\Gamma \not\ni -I$ ($\chi = id.$)

Let Γ be a fuchsian group of the first kind not containing the element $-I$ and suppose that Γ has a non-compact fundamental domain. And we assume that the number of regular cusps of Γ is two. If $\chi = id.$, then we write $\zeta_{\Gamma,\chi}(s) = \zeta_\Gamma(s)$. The notation and assumptions being as above, d_1 is given by the following

$$d_1 = \frac{1}{2} \operatorname*{Res}_{s=0} \zeta_\Gamma(s).$$

Here we give two remarks:

1) For a general discontinuous group Γ, we can obtain a similar result.
2) The number d_1 has another representation (Christian):

$$d_1 = \frac{1}{2} \operatorname*{Res}_{s=\frac{1}{2}} \frac{Z'_\Gamma}{Z_\Gamma}(s)$$

where $Z_\Gamma(s)$ denotes the *Selberg zeta-function* for Γ, i.e.

$$Z_\Gamma(s) = \prod_\alpha \prod_{n=0}^{\infty} (1 - N(P_\alpha)^{-s-n}), \quad \operatorname{Re} s > 1,$$

here $\{P_\alpha\}$ denotes a complete system of representatives of the primitive hyperbolic conjugacy classes in Γ.

(2) The case 2: $\Gamma \ni -I$

Let Γ be a fuchsian group of the first kind containing the element $-I$ and suppose that Γ has a non-compact fundamental domain. We also suppose that Γ is reduced at infinity. Let χ be an odd character such that $\chi^2 \neq 1$ and $\chi(\begin{pmatrix} 1 & 1 \\ 0 & 1 \end{pmatrix}) = 1$. Then d_1 is given by

$$d_1 = \frac{1}{2} \sum_{\{M\}} \frac{\chi(M)}{[\Gamma(M) : \pm I]} \frac{\overline{\zeta}}{1 - \overline{\zeta}^2} + \frac{1}{2} \operatorname*{Res}_{s=0} \zeta_{\Gamma,\chi}(s) - \frac{1}{4} \Psi_\chi\left(\frac{1}{2}\right),$$

where the sum is taken over the distinct elliptic conjugacy classes in $\Gamma/\{\pm I\}$, $\overline{\zeta}$ is one of the eigenvalues of M, $\Gamma(M)$ denotes the centralizer of M in Γ, and $\Psi_\chi(s)$ denotes the function defined by

$$\Psi_\chi(s) = -\sqrt{-1}\sqrt{\pi} \frac{\Gamma(s)}{\Gamma(s + \frac{1}{2})} \sum_{\substack{c>0,\ d \bmod c \\ \begin{pmatrix} * & * \\ c & d \end{pmatrix} \in \Gamma}} \frac{\overline{\chi}(c,d)}{|c|^{2s}},$$

where $\Psi_\chi\left(\frac{1}{2}\right) = \pm 1$.

Remark. Our assumptions on Γ and χ are not essential and convenient ones. For the general case, we obtain the contribution from parabolic part to d_1 in the same way as in the above case.

(3) The case of $\Gamma = \Gamma_0(p)$ (p: prime) Let p be a prime number such that $p \equiv 3 \bmod 4$, $p \neq 3$. Let $\Phi_0(p)$ be the group generated by the group $\Gamma_0(p)$ and $\kappa = \begin{pmatrix} 0 & -\sqrt{p}^{-1} \\ \sqrt{p} & 0 \end{pmatrix}$. We put

$$\chi(L) = \left(\frac{d}{p}\right) \text{ for } L = \begin{pmatrix} a & b \\ c & d \end{pmatrix} \in \Gamma_0(p).$$

Since $\chi(\kappa^2) = \chi(-I) = -1$, we can defined the odd characters $\chi^\pm$ on $\Phi_0(p)$ such that $\chi^\pm(\kappa) = \pm\sqrt{-1}$. Then we have

$$S_1(\Gamma_0(p), \chi) = S_1(\Phi_0(p), \chi^+) \otimes S_1(\Phi_0(p), \chi^-).$$

Now we put $d^{\pm 1} = \dim S_1(\Phi_0(p), \chi^\pm)$. Then we have

$$d_1 = \dim S_1(\Gamma_0(p), \chi) = d_1^+ + d_1^-.$$

By applying the result of the case 2 to $d_1^\pm$, we have the following

$$d_1 = \frac{1}{2} \operatorname*{Res}_{\delta=0} Z^*(\delta), \qquad (*)$$

$$Z^*(\delta) = \sum_{\alpha=1}^{\infty} \sum_{k=1}^{\infty} \frac{\chi(P_\alpha)^k \log \lambda_{0,\alpha}}{\lambda_{0,\alpha}^k - \lambda_{0,\alpha}^{-k}} \left|\lambda_{0,\alpha}^k + \lambda_{0,\alpha}^{-k}\right|^{-\delta},$$

$$(= \zeta_{\Gamma,\chi}(\delta))$$

where $\lambda_{0,\alpha}$ denotes the eigenvalue (> 1) of representative P_α of the hyperbolic conjugacy classes $\{P_\alpha\}$ in $\Gamma_0(p)/\{\pm I\}$.

Remark. 1) H. Petersson obtained the difference

$$d_1^- - d_1^+ = \frac{1}{2}(h-1)$$

by the Riemann-Roch theorem, where h denotes the class number of $\mathbb{Q}(\sqrt{-p})$. This is the dihedral part of d_1.

2) Combining the above result $(*)$ with Serre's result, we have the following remarkable equality

$$\operatorname*{Res}_{\delta=0} Z^*(\delta) = (h-1) + 4(s+2a),$$

where s (resp. a) is the number of the normal closure of a quartic (resp. non-real quintic) fields with discriminant $-p$ (resp. p^2) whose associated representations satisfy a certain condition. These numbers s and a are very complicated. For further details, see the following:

J.-P. Serre, Modular forms of weight one and Galois representations, in Algebraic Number Fields (Ed. Fröhlich), Academic Press, 1977.

§ 3. The Selberg eigenspace

Let S denote the complex upper half-plane and we put $G = \mathrm{SL}(2, \mathbb{R})$. Consider direct products

$$\widetilde{S} = S \times T, \quad \widetilde{G} = G \times T,$$

where T denotes the real torus. The operation of an element (g, α) of $\widetilde{G}$ on $\widetilde{S}$ is represented as follows:

$$\widetilde{S} \ni (z, \phi) \longrightarrow (g, \alpha)(z, \phi) = \left(\frac{az + b}{cz + d}, \ \phi + \arg(cz + d) - \alpha \right) \in \widetilde{S},$$

where $g = \left(\begin{smallmatrix} a & b \\ c & d \end{smallmatrix}\right) \in G$. The $\widetilde{G}$-invariant measure is given by

$$d(z, \phi) = d(x, y, \phi) = \frac{dx \wedge dy \wedge d\phi}{y^2}.$$

By the correspondence $G \ni g \longleftrightarrow (g, 0) \in \widetilde{G}$, we identify the group G with a subgroup $G \times \{0\}$ of $\widetilde{G}$, and so a subgroup Γ of G identify with a subgroup $\Gamma \times \{0\}$ of $\widetilde{G}$. We define a mapping $T_{(g,\alpha)}$ by

$$(T_{(g,\alpha)} f)(z, \phi) = f\left((g, \alpha)(z, \phi) \right).$$

For an element $g \in G$, we put $T_{(g,0)} = T_g$. Then we have

$$(T_g f)(z, \phi) = f\left(\frac{az + b}{cz + d}, \ \phi + \arg(cz + d) \right), \quad g = \begin{pmatrix} a & b \\ c & d \end{pmatrix}.$$

Let Γ be a fuchsian group of the first kind that does not contain the element $-I$. We denote by $\mathfrak{M}_\Gamma(k, \lambda) = \mathfrak{M}(k, \lambda)$ the set of all functions $f(z, \phi)$ satisfying the following conditions:

(i) $f(z, \phi) \in L^2(\Gamma \backslash \widetilde{S})$, i.e. $f(z, \phi)$ is a measurable function on $\widetilde{S}$ such that

$$(T_\gamma f)(z, \phi) = f(z, \phi) \text{ for } \gamma \in \Gamma,$$

and $\displaystyle\int_{\Gamma \backslash \widetilde{S}} |f(z, \phi)|^2 \, d(z, \phi) < \infty;$

(ii)

$$\widetilde{\Delta} f = \lambda f, \quad \frac{\partial}{\partial \phi} f(z,\phi) = -\sqrt{-1}k\, f(z,\phi),$$

where

$$\widetilde{\Delta} = y^2\left(\frac{\partial^2}{\partial^2 x^2} + \frac{\partial^2}{\partial^2 y^2}\right) + \frac{5}{4}\frac{\partial^2}{\partial \phi^2} + y\frac{\partial}{\partial \phi}\frac{\partial}{\partial x},$$

which we call the Casimir operator of $\widetilde{S}$. We call $\mathfrak{M}(k,\lambda)$ the Selberg eigenspace of Γ. Then the following equality holds:

Lemma. $\mathfrak{M}(1,-\frac{3}{2}) = \left\{ e^{-\sqrt{-1}} y^{\frac{1}{2}} F(z) : F(z) \in S_1(\Gamma) \right\}$, *and hence*

$$d_1 = \dim \mathfrak{M}\left(1, -\frac{3}{2}\right).$$

Here we give two remarks.

1) d_1 corresponds to multiplicity of the limit of discrete series for $\mathrm{SL}(2,\mathbb{R})$.

2) For $k \geq 1$,

$$\mathfrak{M}(k, -k(k+\tfrac{1}{2})) = \left\{ e^{-\sqrt{-1}k} y^{\frac{k}{2}} F(z) : F(z) \in S_k(\Gamma) \right\}.$$

§ 4. The compact case

In this section we suppose that the group Γ has a compact fundamental domain in S. It is well known that every eigenspace $\mathfrak{M}(k,\lambda)$ defined in Section 3 is finite dimensional and orthogonal to each other, and also the eigenspaces span together the space $L^2(\Gamma\backslash\widetilde{S})$. We put $\boldsymbol{\lambda} = (k,\lambda)$. For every invariant integral operator with a kernel function $k(z,\phi;z',\phi')$ on $\mathfrak{M}(k,\lambda)$, we have

$$\int_{\widetilde{S}} k(z,\phi;z',\phi')\, f(z',\phi')\, d(z',\phi') = h(\boldsymbol{\lambda}) f(z,\phi)$$

for $f \in \mathfrak{M}(k,\lambda)$, where $h(\boldsymbol{\lambda})$ does not depend on f so long as f is in $\mathfrak{M}(k,\lambda)$. We know that there is a basis $\{f^{(n)}\}_{n=1}^{\infty}$ of the space $L^2(\Gamma\backslash\widetilde{S})$ such that each $f^{(n)}$ satisfies the condition (ii) in Section 3. Then we put $\boldsymbol{\lambda}^{(n)}$ for such a spectra. Now we obtain the following Selberg trace formula for $L^2(\Gamma\backslash\widetilde{S})$:

$$\sum_{n=1}^{\infty} h(\boldsymbol{\lambda}^{(n)}) = \sum_{M\in\Gamma} \int_{\widetilde{D}} k(z,\phi;M(z,\phi))\, d(z,\phi), \tag{1}$$

where $\widetilde{D}$ denotes a compact fundamental domain of Γ in $\widetilde{S}$ and $k(z,\phi;z',\phi')$ is an invariant kernel of (a)–(b) type in the sense of Selberg such that the series on the left-hand side of (1) is absolutely convergent. Denote by $\Gamma(M)$ the centralizer of M in Γ and put $\widetilde{D}_M = \Gamma(M)\backslash\widetilde{S}$. Then the right-hand side of (1) equals

$$\sum_\ell \int_{\widetilde{D}_{M_\ell}} k(z,\phi; M_\ell(z,\phi))\, d(z,\phi), \tag{2}$$

where the sum is taken over the distinct conjugacy classes of Γ.

We consider an invariant integral operator on the Selberg eigenspace $\mathfrak{M}(k,\lambda)$ defined by

$$\omega_\delta(z,\phi;z',\phi') = \left|\frac{(yy')^{1/2}}{(z-z')/2\sqrt{-1}}\right|^\delta \frac{(yy')^{1/2}}{(z-\overline{z}')/2\sqrt{-1}} e^{-\sqrt{-1}(\phi-\phi')}.$$

It is easy to see that our kernel ω_δ is an invariant kernel of (a)–(b) type under the condition $\delta > 1$, and vanishes on $\mathfrak{M}(k,\lambda)$ for all $k \neq 1$. Since $\widetilde{D}$ is compact, the distribution of spectra (k,λ) is discrete and

$$(1,\mu_\beta), \quad \mu_\beta < 0, \quad \mu_\beta \neq -\frac{3}{2}, -\frac{1}{2},$$
$$\left(1,-\frac{3}{2}\right) \left(1,-\frac{1}{2}\right)$$

gives the complete set of spectra of type $(1,*)$ (Kuga). But the spectra of types $-\frac{3}{2} < \mu_\beta < 0$ in $(1,\mu_\beta)$ do not appear actually in the complete set (Bargmann)[1]. Then we put

$$\mu_1 = -\frac{3}{2},\ \mu_2,\ \mu_3,\ldots; \quad d_\beta = \dim \mathfrak{M}(1,\mu_\beta), \quad \beta = 1,2,3,\ldots.$$

Hence the left-hand side of the trace formula (1) is equal to $\sum_{\beta=1}^\infty d_\beta \Lambda_\beta$, where Λ_β denotes the eigenvalue of ω_δ in $\mathfrak{M}(1,\mu_\beta)$. For the eigenvalue Λ_β, using the special eigenfunction

$$f(z,\phi) = e^{\sqrt{-1}\phi} y^{v_\beta}, \quad \mu_\beta = v_\beta(v_\beta - 1) - \frac{5}{4}$$

for a spectrum $(1,\mu_\beta)$, we obtain

$$\Lambda_\beta = 2^{2+\delta}\pi \frac{\Gamma(\frac{1}{2})\Gamma(\frac{1+\delta}{2})}{\Gamma(\delta)\Gamma(1+\frac{\delta}{2})} \Gamma\left(\frac{\delta}{2} + \sqrt{-1}r_\beta\right) \Gamma\left(\frac{\delta}{2} - \sqrt{-1}r_\beta\right),$$

where $v_\beta = \frac{1}{2} + \sqrt{-1}r_\beta$. In general, it is known that the series $\sum_{\beta=1}^\infty d_\beta\Lambda_\beta$ is absolutely convergent for $\delta > 1$. By the Stirling formula, we see that the series above is also absolutely and uniformly convergent for all bounded δ except $\delta = \pm(2v_\beta - 1)$. We remark that

$$\delta = 0 \Longleftrightarrow v_\beta = \frac{1}{2} \Longleftrightarrow \mu_\beta = -\frac{3}{2} \Longleftrightarrow \beta = 1.$$

We shall calculate the components of trace appearing in (2).

[1]This remark was informed by Professor Satake to the author.

[1] $M = I$, $\omega_\delta(z,\phi; I(z,\phi)) = 1$.

$$J(I) = \int_{\widetilde{D}_I} d(z,\phi) = \int_{\widetilde{D}} d(z,\phi) < \infty.$$

[2] Hyperbolic conjugacy classes

For the primitive hyperbolic P, we put

$$g^{-1}Pg = \begin{pmatrix} \lambda_0 & 0 \\ 0 & \lambda_0^{-1} \end{pmatrix}, \; g \in G, \; |\lambda_0| > 1$$

and $\Gamma' = g^{-1}\Gamma g$. Then

$$\Gamma'\left(\begin{pmatrix} \lambda_0 & 0 \\ 0 & \lambda_0^{-1} \end{pmatrix}\right) = g^{-1}\Gamma(P)g.$$

The hyperbolic component is calculated as follows:

$$\begin{aligned} J(P^k) &= \int_{\widetilde{D}_P} \omega_\delta(\,z,\phi;\; P^k(z,\phi)\,)\, d(z,\phi) \\ &= \int_{g^{-1}\widetilde{D}_P} \omega_\delta(\,g(z,\phi);\; P^k g(z,\phi)\,)\, d(z,\phi) \\ &= \int_{g^{-1}\widetilde{D}_P} \omega_\delta(\,z,\phi;\; g^{-1}P^k g(z,\phi)\,)\, d(z,\phi) \end{aligned}$$

$$= 2\pi(2^{\delta+1}\sqrt{-1})|\lambda_0^k|^{\delta+1}(\mathrm{sgn}\lambda_0)^k \int_{g^{-1}D_P} \frac{y^{\delta-1}}{(z-\lambda_0^{2k}\overline{z})|z-\lambda_0^{2k}\overline{z}|^\delta} dxdy$$

$$\left(e^{\sqrt{-1}\arg\lambda_0^{-k}} = (\mathrm{sgn}\lambda_0)^k\right)$$

$$= (2^{3+\delta}\pi)\frac{\Gamma(\frac{1}{2})\Gamma(\frac{\delta+1}{2})}{\Gamma(\frac{\delta+2}{2})} \cdot \frac{(\mathrm{sgn}\lambda_0)^k \log|\lambda_0|}{|\lambda_0^{-k}-\lambda_0^k||\lambda_0^{-k}+\lambda_0^k|}.$$

Let $\{P_\alpha\}$ be a complete system of representatives of the primitive hyperbolic conjugacy classes in Γ and let $\lambda_{0,\alpha}$ be the eigenvalue ($|\lambda_{0,\alpha}| > 1$) of P_α. Then, the hyperbolic component $J(P)$ can be expressed as follows:

$$\begin{aligned} J(P) &= \sum_{\alpha=1}^{\infty}\sum_{k=1}^{\infty} J(P_\alpha^k) \\ &= \frac{2^{3+\delta}\pi^{\frac{3}{2}}\Gamma(\frac{\delta+1}{2})}{\Gamma(\frac{\delta+2}{2})} \sum_{\alpha=1}^{\infty}\sum_{k=1}^{\infty} \frac{(\mathrm{sgn}\lambda_{0,\alpha})^k \log|\lambda_{0,\alpha}|}{|\lambda_{0,\alpha}^k - \lambda_{0,\alpha}^{-k}|} |\lambda_{0,\alpha}^k + \lambda_{0,\alpha}^{-k}|^{-\delta}. \end{aligned}$$

3 Elliptic conjugacy classes

Let ρ, $\overline{\rho}$ be the fixed points of an elliptic element R ($\rho \in S$) and ζ, $\overline{\zeta}$ be the eigenvalues of R. We denote by Φ a linear transformation such that maps S into a unit circle:

$$w = \Phi(z) = \frac{z-\rho}{z-\overline{\rho}}.$$

Then we have $\Phi R \Phi^{-1} = \left(\begin{smallmatrix} \zeta & 0 \\ 0 & \overline{\zeta} \end{smallmatrix}\right)$ and

$$\frac{Rz-\rho}{Rz-\overline{\rho}} = \frac{\zeta}{\overline{\zeta}} \frac{\zeta-\rho}{\zeta-\overline{\zeta}}.$$

Remark. $R = \left(\begin{smallmatrix} a & b \\ c & d \end{smallmatrix}\right)$. $\operatorname{Im}\zeta > 0$.

$$\text{If } c>0, \text{ then } c\rho + d = \zeta\colon\ \Phi R \Phi^{-1} = \left(\begin{smallmatrix} \overline{\zeta} & 0 \\ 0 & \zeta \end{smallmatrix}\right);$$
$$\text{If } c<0, \text{ then } c\overline{\rho} + d = \zeta\colon\ \Phi R \Phi^{-1} = \left(\begin{smallmatrix} \zeta & 0 \\ 0 & \overline{\zeta} \end{smallmatrix}\right).$$

The elliptic component is calculated as follows:

$$\begin{aligned} J(P) &= \int_{\widetilde{D}_R} \omega_\delta(\,z,\phi\,:\,R(z,\phi)\,)\,d(z,\phi) \\ &= \frac{16\pi^2\overline{\zeta}}{[\Gamma(R):1]} \int_0^1 \frac{(1-r^2)^{\delta-1}}{(1-\overline{\zeta}^2 r^2)|1-\overline{\zeta}^2 r^2|^\delta}\,dr. \end{aligned}$$

Therefore we obtain

$$\lim_{\delta\to 0} \delta\, J(R) = \frac{8\pi^2}{[\Gamma(R):1]} \cdot \frac{\overline{\zeta}}{1-\overline{\zeta}^2}.$$

Since R and R^{-1} are not conjugate to each other and $\dfrac{\overline{\zeta}}{1-\overline{\zeta}^2}$ is pure imaginary, we have

$$\lim_{\delta\to 0} \delta\, J(R) + \lim_{\delta\to 0} \delta\, J(R^{-1}) = 0.$$

As conclusion, the contribution from elliptic classes to d_1 vanishes.

Now we put

$$\zeta_1(\delta) = \sum_{\alpha=1}^{\infty}\sum_{k=1}^{\infty} \frac{(\mathrm{sgn}\lambda_{0,\alpha})^k \log|\lambda_{0,\alpha}|}{|\lambda_{0,\alpha}^k - \lambda_{0,\alpha}^{-k}|} |\lambda_{0,\alpha}^k + \lambda_{0,\alpha}^{-k}|^{-\delta}. \tag{3}$$

Then, by the trace formula (1), the Dirichlet series (3) extends to a (complex) meromorphic function on the whole δ-plane and has a simple pole at

$\delta = 0$. Finally, multiply the both side of (1) by δ and tend δ to zero, then by the above $\boxed{1}$, $\boxed{2}$ and $\boxed{3}$, the limit is expressed as follows:

$$d_1 = \dim S_1(\Gamma) = \dim \mathfrak{M}\left(1, -\frac{3}{2}\right) \tag{4}$$
$$= \frac{1}{2} \operatorname*{Res}_{\delta=0} \zeta_1(\delta).$$

Remark. In this remark we suppose that Γ has a compact fundamental domain in S and Γ contains the element $-I$. Let $S_1(\Gamma, \chi)$ be the linear space of automorphic forms of weight 1 on the group Γ with odd character χ. Then we have the following dimension formula in the same way as in the case $\Gamma \ni -I$.

$$d_1 = \dim S_1(\Gamma, \chi) = \frac{-1}{2} \sum_{\{M\}} \frac{\chi(M)}{[\Gamma(M) : \pm I]} \frac{\overline{\zeta}}{1 - \overline{\zeta}^2} + \frac{1}{2} \operatorname*{Res}_{\delta=0} \zeta_2(\delta), \tag{5}$$

where the sum is taken over the distinct elliptic conjugacy classes of $\Gamma/\{\pm I\}$, $\Gamma(M)$ denotes the centralizer of M in Γ, $\overline{\zeta}$ is one of the eigenvalues of M, and $\zeta_2(\delta)$ denotes the Selberg type zeta-function defined by

$$\zeta_2(\delta) = \sum_{\alpha=1}^{\infty} \sum_{\beta=1}^{\infty} \frac{\chi(P_\alpha)^k \log \lambda_{0,\alpha}}{\lambda_{0,\alpha}^k - \lambda_{0,\alpha}^{-k}} |\lambda_{0,\alpha}^k + \lambda_{0,\alpha}^{-k}|^{-\delta}. \tag{6}$$

Here $\lambda_{0,\alpha}$ denotes the eigenvalue ($\lambda_{0,\alpha} > 1$) of representative P_α of the primitive hyperbolic conjugacy classes $\{P_\alpha\}$ in $\Gamma/\{\pm I\}$.

§ 5. The finite case 1: $\Gamma \not\ni -I$

Let Γ be a fuchsian group of the first kind not containing the element $-I$, and suppose that Γ has a non-compact fundamental domain $\widetilde{D}$ in $\widetilde{S}$. Then, we see that the integral

$$\int_{\widetilde{D}} \sum_{M \in \Gamma} \omega_\delta(\, z, \phi, \, ; \, M(z, \phi)\,)\, d(z, \phi)$$

is uniformly bounded on a neighborhood of each irregular cusp of Γ, and that the number of regular cusps of Γ is even. So we can assume that κ_1, κ_2 is a maximal set of cusps of Γ which are regular and not equivalent in Γ. Let Γ_i be the stabilizer in Γ of κ_i, and fix an element $\sigma_i \in G$ such that $\sigma_i \infty = \kappa_i$ and such that $\sigma_i^{-1} \Gamma \sigma_i$ is equal to the group $\left\{ \left(\begin{smallmatrix} 1 & m \\ 0 & 1 \end{smallmatrix}\right) : m \in \mathbb{Z} \right\}$. Then the Eisenstein series attached to the regular cusp κ_i is defined by

$$E_i(z, \phi; s) = \sum_{\substack{\sigma \in \Gamma_i \backslash \Gamma \\ \sigma_i^{-1}\sigma = \left(\begin{smallmatrix} * & * \\ c & d \end{smallmatrix}\right)}} \frac{y^s}{|cz + d|^{2s}} e^{-\sqrt{-1}(\,\phi + \arg(cz+d)\,)} \quad (i = 1, 2), \tag{7}$$

where $s = \sigma + \sqrt{-1}r$ with $\sigma > 1$. The series (7) has Fourier expansion at κ_j in the form

$$E_i(\sigma_j(z,\phi)\,;\, s) = \sum_{m=-\infty}^{\infty} a_{i,j,m}(y,\phi\,;\, s)\, e^{2\pi\sqrt{-1}mx}.$$

The constant term $a_{i,j,0}(y,\phi\,;\,s)$ is given by

$$\begin{aligned} e^{\sqrt{-1}\phi} a_{i,j,0}(y,\phi\,;\,s) &= a_{i,j,0}(y\,;\,s) \\ &= \delta_{ij} y^s + \psi_{ij}(s) y^{1-s} \end{aligned}$$

with

$$\psi_{ij}(s) = -\sqrt{-1}\sqrt{\pi}\frac{\Gamma(s)}{\Gamma(s+\frac{1}{2})}\sum_{c\neq 0}\frac{(\text{sgn } c)\cdot N_{ij}(c)}{|c|^{2s}},$$

where $N_{ij}(c) = \#\left\{0 \leqq d < |c| \,:\, \left(\begin{smallmatrix} * & * \\ c & d\end{smallmatrix}\right) \in \sigma_i^{-1}\Gamma\sigma_i\right\}$. We put

$$\Phi(s) = (\psi_{ij}(s)).$$

Then, we know

$$N_{ij}(c) = \left\{\Gamma_\infty \tau \Gamma_\infty \,:\, \tau \in \sigma_i^{-1}\Gamma\sigma_i,\ c(\tau) = c\right\}.$$

Therefore, $N_{ij}(-c) = N_{ji}(c)$ and hence $\psi_{ij}(s) = -\psi_{ji}(s)$, i.e. the Eisenstein matrix $\Phi(s)$ is a skew-symmetric matrix.

Since Γ is of finite type, the integral operator defined by ω_δ is not generally completely continuous on $L^2(\Gamma\backslash\widetilde{S})$ and the space $L^2(\Gamma\backslash\widetilde{S})$ admits the following spectral decomposition

$$L^2(\Gamma\backslash\widetilde{S}) = L_0^2(\Gamma\backslash\widetilde{S}) \oplus L_{\text{sp}}^2(\Gamma\backslash\widetilde{S}) \oplus L_{\text{cont}}^2(\Gamma\backslash\widetilde{S}),$$

where L_0^2 is the space of cusp forms and is discrete, L_{sp}^2 is the discrete part of the orthogonal complement of L_0^2 and L_{cont}^2 is continuous part of the spectra. We put

$$\begin{aligned} &\widetilde{H}_\delta(z,\phi\,;\,z',\phi') \\ &= \frac{1}{8\pi^2}\sum_{i=1}^{2}\int_{-\infty}^{\infty} h(r) E_i(z,\phi\,;\,\frac{1}{2}+\sqrt{-1}r)\overline{E_i(z',\phi'\,;\,\frac{1}{2}+\sqrt{-1}r)}\,dr. \end{aligned}$$

Here $h(r)$ denotes the eigenvalue of ω_δ in $\mathfrak{M}(1,\lambda)$ given in Section 4:

$$h(r) = 2^{2+\delta}\pi\frac{\Gamma(\frac{1}{2})\Gamma(\frac{1+\delta}{2})}{\Gamma(\delta)\Gamma(1+\frac{\delta}{2})}\Gamma\left(\frac{\delta}{2}+\sqrt{-1}r\right)\Gamma\left(\frac{\delta}{2}-\sqrt{-1}r\right) \tag{8}$$

with $\lambda = s(s-1) - \frac{5}{4}$ and $s = \frac{1}{2} + \sqrt{-1}r$. We put

$$K_\delta(z,\phi;\, z',\phi') = \sum_{M\in\Gamma} \omega_\delta(z,\phi;\, M(z',\phi'))$$

and $\widetilde{K}_\delta = K_\delta - \widetilde{H}_\delta$. Then the integral operator $\widetilde{K}_\delta$ is now complete continuous on $L^2(\Gamma\backslash\widetilde{S})$ and has all discrete spectra of K_δ. Furthermore, an eigenvalue of $f(z,\phi)$ for $\widetilde{K}_\delta$ in $L_0^2(\Gamma\backslash\widetilde{S}) \oplus L_{\mathrm{sp}}^2(\Gamma\backslash\widetilde{S})$ is equal to that for K_δ and the image of $\widetilde{K}_\delta$ on it is contained in $L^2(\Gamma\backslash\widetilde{S})$.

cf. H. Ishikawa: On the trace formula for Hecke operators,

J. Fac. Sci. Univ. Tokyo, Sec. IA, **20** (1973), 217-238, §4.4

Considering the trace of $\widetilde{K}_\delta$ on $L_0^2(\Gamma\backslash\widetilde{S})$, we now obtain the following modified trace formula

$$\sum_{n=1}^{\infty} h(\boldsymbol{\lambda}^{(n)}) = \int_{\widetilde{D}} \widetilde{K}_\delta(z,\phi;\, z,\phi)\, d(z,\phi),$$

where each $\boldsymbol{\lambda}^{(n)}$ denotes an eigenvalue corresponding to an orthogonal basis $\{f^{(n)}\}$ of $L_0^2(\Gamma\backslash\widetilde{S})$. We put

$$\int_{\widetilde{D}} \widetilde{K}_\delta(z,\phi;\, z,\phi)\, d(z,\phi) = J(I) + J(P) + J(R) + J(\infty),$$

where $J(I)$, $J(P)$, $J(R)$ and $J(\infty)$ denote respectively the identity component, the hyperbolic component, the elliptic component and the parabolic component of the trace. Then the components $J(I)$, $J(P)$ and $J(R)$ are as given in Section 4 and in the following we shall calculate the component $J(\infty)$.

Let $\widetilde{D}_i$ be a fundamental domain of the stabilizer Γ_i of the cusp κ_i in Γ. Then we have

$$J(\infty) = \lim_{Y\to\infty} \left\{ \sum_{i=1}^{2} \int_{\widetilde{D}_i^Y} \sum_{\substack{M\in\Gamma_i \\ M\neq I}} \omega_\delta(z,\phi;\, M(z,\phi))\, d(z,\phi) \right.$$
$$\left. - \frac{1}{8\pi^2} \sum_{i=1}^{2} \int_{\widetilde{F}_i^Y} \int_{-\infty}^{\infty} h(r) E_i(z,\phi;\, \frac{1}{2} + \sqrt{-1}r) \overline{E_i(z,\phi;\, \frac{1}{2} + \sqrt{-1}r)}\, dr \right\},$$

where $\widetilde{D}_i^Y$ denotes the domain consisting of all points (z,ϕ) in $\widetilde{D}_i$ such that $\mathrm{Im}\,(\sigma_i^{-1}z) < Y$, and $\widetilde{F}_i^Y$ the domain consisting of all $(z,\phi) \in \widetilde{D}$ such that $\mathrm{Im}\,(\sigma_i^{-1}z) < Y$ for $i = 1, 2$. Making use of a summation formula due to Euler-MacLaurin and the Maass-Selberg relation, we have the following

$$\int_{\widetilde{D}_i^Y} \sum_{\substack{M\in\Gamma_i \\ M\neq I}} \omega_\delta(z,\phi;\, M(z,\phi))\, d(z,\phi) = 2^2\pi \frac{\Gamma(\frac{1}{2})\Gamma(\frac{\delta+1}{2})}{\Gamma(1+\frac{1}{2})} \log Y + \varepsilon(\delta) + o(1)$$

as $Y \to \infty$, where $\varepsilon(\delta)$ denotes a function of δ such that $\lim_{\delta\to 0}\delta\,\varepsilon(\delta)=0$, and

$$\frac{1}{8\pi^2}\int_{\widetilde{F}_i^Y}\int_{-\infty}^{\infty} h(r)E_i\left(z,\phi;\frac{1}{2}+\sqrt{-1}r\right)\overline{E_i\left(z,\phi;\frac{1}{2}+\sqrt{-1}r\right)}\,dr\,d(z,\phi)$$
$$=2^2\pi\frac{\Gamma(\frac{1}{2})\Gamma(\frac{\delta+1}{2})}{\Gamma\left(1+\frac{1}{2}\right)}\log Y-\frac{1}{4\pi}\int_{-\infty}^{\infty}h(r)\frac{\psi'_{ij}}{\psi_{ij}}(\frac{1}{2}+\sqrt{-1}r)\,dr+o(1)$$

as $Y\to\infty$ $(j\neq i)$. By the expression (8) of $h(r)$, we have

$$h(r)=O\left(\frac{|r|^\delta}{|r|e^{\pi|r|}}\right), \tag{9}$$

and the operator $\widetilde{K}_\delta$ is complete continuous on $L^2(\Gamma\backslash\widetilde{S})$. Therefore we have that

$$\lim_{\delta\to 0}\delta\int_{-\infty}^{\infty}h(r)\frac{\psi'_{ij}}{\psi_{ij}}\left(\frac{1}{2}+\sqrt{-1}r\right)\,dr=0.$$

It is now clear that the above result, with combined with the formula (4), proves the following

Theorem 1. *Let Γ be a fuchsian group of the first kind not containing the element $-I$ and suppose that the number of regular cusps of Γ is two. Then the dimension d_1 for the space consisting of all cusp forms of weight* 1 *with respect to Γ is given by*

$$d_1=\frac{1}{2}\operatorname*{Res}_{\delta=0}\zeta_1(\delta), \tag{10}$$

where $\zeta_1(\delta)$ denotes the Selberg type zeta-function defined by (3) *in Section* 4.

Remark. For a general case, we can obtain a similar result. Let Γ be a general discontinuous group of finite type not containing the element $-I$. Then we can prove that in the same way as in the above case, the contribution from parabolic classes to d_1 vanishes.

§ 6. The finite case 2: $\Gamma \ni -I$

Let Γ be a fuchsian group of the first kind and assume that Γ contains the element $-I$ and has a non-compact fundamental domain $\widetilde{D}$ in the space $\widetilde{S}$. Let χ be a unitary representation of Γ of degree 1 such that $\chi(-I)=-1$. We denote by $S_1(\Gamma,\chi)$ the linear space of cusp forms of weight 1 on the group Γ with the odd character χ and by d_1 the dimension of the space

$S_1(\Gamma, \chi)$. In this section we shall give similar formula of d_1 when Γ is of finite type reduced at infinity and $\chi^2 \neq 1$.

Since Γ is of finite type reduced at ∞, ∞ is a cusp of Γ and the stabilizer Γ_∞ of ∞ in Γ is equal to $\pm\Gamma_0$ with $\Gamma_0 = \{\left(\begin{smallmatrix} 1 & m \\ 0 & 1 \end{smallmatrix}\right) : m \in \mathbb{Z}\}$. The Eisenstein series $E_\chi(z, \phi; s)$ attached to ∞ and χ is then defined by

$$E_\chi(z, \phi; s) = \sum_{\substack{M \in \Gamma_\infty \backslash \Gamma \\ M = \left(\begin{smallmatrix} * & * \\ c & d \end{smallmatrix}\right)}} \frac{\overline{\chi}(M) y^s}{|cz + d|^{2s}} e^{-\sqrt{-1}(\phi + \arg(cz+d))}, \tag{11}$$

where $s = \sigma + \sqrt{-1}r$ with $r > 1$. The constant term in the Fourier expansion of (11) at ∞ is given by

$$a_0(y, \phi; s) = e^{-\sqrt{-1}\phi} \left(y^s + \Psi_\chi(s) y^{1-s}\right),$$

where

$$\Psi_\chi(s) = -\sqrt{-1}\sqrt{\pi} \frac{\Gamma(s)}{\Gamma(s + \frac{1}{2})} \sum_{\substack{c>0, d \bmod c \\ \left(\begin{smallmatrix} * & * \\ c & d \end{smallmatrix}\right) \in \Gamma}} \frac{\overline{\chi}(c, d)}{|c|^{2s}}.$$

In the following we only consider the case that

$$\chi\left(\begin{pmatrix} 1 & 1 \\ 0 & 1 \end{pmatrix}\right) = 1,$$

namely χ is singular. Then the parabolic component $J(\infty)$ in the trace formula is given by

$$\begin{aligned} J(\infty) = \lim_{Y \to \infty} \Bigg\{ &\int_0^Y \int_0^1 \int_0^\pi 2 \sum_{\substack{M \in \Gamma \\ M \neq I}} \omega_\delta(z, \phi; M(z, \phi))\, d(z, \phi) \\ &- \int_{\widetilde{F}^Y} \widetilde{H}_\delta(z, \phi; z, \phi)\, d(z, \phi) \Bigg\} \\ = -\frac{1}{4\pi} &\int_{-\infty}^{\infty} h(r) \frac{\Psi'_\chi(\frac{1}{2} + \sqrt{-1}r)}{\Psi_\chi(\frac{1}{2} + \sqrt{-1}r)}\, dr - \frac{1}{4} h(0) \Psi_\chi(\frac{1}{2}) + \varepsilon(\delta) \end{aligned}$$

as $\lim_{\delta \to 0} \delta\, \varepsilon(\delta) = 0$. When we combine this with the formula (5), we are led to the following theorem:

Theorem 2. *Let Γ be a function group of the first kind containing the element $-I$ and suppose that Γ is reduced at infinity. Let χ be a one-dimensional unitary representation of Γ such that $\chi(-I) = -1$, $\chi^2 \neq 1$ and $\chi(\left(\begin{smallmatrix} 1 & 1 \\ 0 & 1 \end{smallmatrix}\right)) = 1$. Then d_1 is given by*

$$d_1 = \dim S_1(\Gamma, \chi)$$

$$= \frac{1}{2} \sum_{\{M\}} \frac{\chi(M)}{[\Gamma(M) : \pm I]} \cdot \frac{\overline{\zeta}}{1 - \overline{\zeta}^2} + \frac{1}{2} \operatorname*{Res}_{\delta=0} \zeta_2(\delta) - \frac{1}{4} \Psi_\chi \left(\frac{1}{2} \right), \tag{12}$$

where the sum is taken over the distinct elliptic conjugacy classes of $\Gamma/\{\pm I\}$, $\Gamma(M)$ denotes the centralizer of M in Γ, $\overline{\zeta}$ is one of eigenvalues of M, and $\zeta_2(\delta)$ denotes the Selberg type zeta-function defined by (6) in Section 4.

We may call the formulas (10) and (12) a kind of Riemann-Roch type theorem for automorphic forms of weight 1 respectively.

Remark. For a general discontinuous group Γ of finite type containing the element $-I$, we obtain the contribution from parabolic classes to d_1 in the same way as in the case of reduced at ∞.

§ 7. The case of $\Gamma_0(p)$

Let p be a prime number such that $p \equiv 3 \bmod 4$, $p \neq 3$ and let $\Phi_0(p)$ be the group generated by the Hecke congruence subgroup $\Gamma_0(p)$ and the element $\kappa = \begin{pmatrix} 0 & -\sqrt{p}^{-1} \\ \sqrt{p} & 0 \end{pmatrix}$. We put

$$\chi(L) = \left(\frac{d}{p} \right) \text{ for } L = \begin{pmatrix} a & b \\ c & d \end{pmatrix} \in \Gamma_0(p).$$

Since $\chi(\kappa^2) = \chi(-I) = -1$, we can define the odd characters $\chi^{\pm}$ on $\Phi_0(p)$ such that $\chi^{\pm}(\kappa) = \pm\sqrt{-1}$. Then we have

$$S_1(\Gamma_0(p), \chi) = S_1(\Phi_0(p), \chi^+) \oplus S_1(\Phi_0(p), \chi^-).$$

We put

$$d_1^{\pm} = \dim S_1(\Phi_0(p), \chi^{\pm}).$$

Then

$$\dim S_1(\Phi_0(p), \chi) = d_1 = d_1^+ + d_1^- .$$

If $\sigma(p)$ is the number of parabolic classes in $\Gamma_0(p)$, then $\sigma(p) = 2$; and if $e_2(p)$, $e_3(p)$ are the number of elliptic classes of order 2, 3 respectively of $\Gamma_0(p)$, then $e_2(p) = 0$, $e_3(p) = 1 + \left(\frac{p}{3}\right)$. Let $\sigma^*(p)$, $e_2^*(p)$, $e_3^*(p)$ denote respectively the number of parabolic classes, the number of elliptic classes of order 2, the elliptic classes of order 3 for $\Phi_0(p)$. Then we have

$$\sigma^*(p) = \frac{1}{2}\sigma(p) = 1,$$

$$e_3^*(p) = \frac{1}{2} e_3(p) = \frac{1}{2}\left(1 + \left(\frac{p}{3}\right)\right),$$
$$e_2^*(p) = \frac{1}{2} e_2(p) + e_2'(p) = e_2'(p),$$

where $e_2'(p)$ denotes the number of elliptic classes of order 2 for $\kappa\Gamma_0(p)$. Moreover it is known that

$$e_2'(p) = (3 - \left(\frac{2}{p}\right))h = \begin{cases} 4h \text{ if } p \equiv 3 \bmod 8, \\ 2h \text{ if } p \equiv 7 \bmod 8, \end{cases}$$

where h denotes the class number of $\mathbb{Q}(\sqrt{-p})$ and this is an odd number. After some calculations, we have the following contribution from elliptic classes to $d_1^{\pm}$:

$$\frac{1}{2} \sum_M \frac{\chi^{\pm}(M)}{[\Gamma(M) : \pm I]} \cdot \frac{\overline{\zeta}}{1 - \overline{\zeta}^2} = \mp \frac{1}{4} h.$$

We also have $\Psi_{\chi^{\pm}}\left(\frac{1}{2}\right) = \mp 1$. Let $\{P_\alpha\}$ be a complete system of representatives of the primitive hyperbolic conjugacy classes in $\Gamma_0(p)$ and let $\lambda_{0,\alpha}$ be the eigenvalue ($\lambda_{0,\alpha} > 1$) of P_α. We put

$$Z^*(\delta) = \sum_{\alpha=1}^{\infty} \sum_{\beta=1}^{\infty} \frac{\chi(P_\alpha) \log \lambda_{0,\alpha}}{|\lambda_{0,\alpha}^k - \lambda_{0,\alpha}^{-k}|} \left|\lambda_{0,\alpha}^k + \lambda_{0,\alpha}^{-k}\right|^{-\delta}.$$

As a consequence, we have the following

$$d_1 = d_1^+ + d_1^- = \frac{1}{2} \operatorname*{Res}_{\delta=0} Z^*(\delta).$$

Remark. Finally I will present some additional problems.

(1) The first one is extension of our results to the Hilbert modular case (two-variable case).

(2) Is there the finite closed formula of d_1 by fundamental properties of Γ? It is a problem very difficult to answer at the present stage of our investigations. But the following problem will be possible to solve it:

$$d_1 \neq 0 \underset{?}{\Longleftrightarrow} \text{special value of the Selberg zeta-function,}$$
$$\text{i.e. } Z_{\Gamma,\chi}\left(\frac{1}{2}\right) = 0.$$

(3) Algebraicity of $\operatorname*{Res}_{s=0} \zeta_i(s)$ $(i = 1, 2)$.

(4) Determine the behavior of d_1 by deformations of Γ.

§ 8. Trace of Hecke operators in the case of weight 1

First we set the following notation:

$G = \mathrm{SL}(2, \mathbb{R})$,

$G \ni \alpha$ such that $\alpha \Gamma \alpha^{-1}$ is commensurable with Γ,

$\Gamma' = \langle \Gamma, \alpha \rangle$,

χ: unitary representation of Γ' of finite degree s.t.
$[\Gamma : \Gamma_\chi] < \infty$ (Γ_χ: the kernel of χ in Γ),

$k \in \mathbb{Z}^+$,

$\Gamma \alpha \Gamma = \bigcup_\mu M_\mu \Gamma$: right Γ-coset decomposition of $\Gamma \alpha \Gamma$.

Then, the Hecke operator $T(\Gamma \alpha \Gamma)$ acts on $S_k(\Gamma, \chi)$ define by the following

$$(T(\Gamma \alpha \Gamma))(z) = \sum_\mu \chi(M_\mu) F(M_\mu^{-1} z)(c_\mu z + d_\mu)^{-k},$$

where $F(z) \in S_1(\Gamma, \chi)$, $M_\mu^{-1} = \begin{pmatrix} * & * \\ c_\mu & d_\mu \end{pmatrix}$.

Now we propose that

'tr$T(\Gamma \alpha \Gamma)$ in the case $k = 1$?'

This problem was brought forward by Eichler in 1958 (notes of the lecture in Japan taken by Y. Taniyama).

In the following we present a formula of tr$T(\Gamma \alpha \Gamma)$ in the case $k = 1$ by using the Selberg trace formula. We shall only state the result in the case $\Gamma = \Gamma_0(p)$ (p : prime).[2]

Theorem 3. *The trace t_1 of $T(\Gamma \alpha \Gamma)$ on $S_1(\Gamma_0(p), \chi)$ is given by*

$$t_1 = \frac{1}{2} \operatorname*{Res}_{\delta=0} H(\delta).$$

Here $H(\delta)$ denotes the Selberg type zeta-function defined by

$$H(\delta) = \sum_{[g] \in \mathfrak{S}} \frac{\mathrm{tr}\chi(g) \log \lambda_0}{|\lambda + \lambda^{-1}|^\delta |\lambda - \lambda^{-1}|}$$

where $\mathfrak{S}$ denotes a complete system of hyperbolic conjugacy classes in $\Gamma \alpha \Gamma$ which leave parabolic elements of Γ fixed, λ an eigenvalue of $g \in [g] \in \mathfrak{S}$ and λ_0 the eigenvalue ($\lambda_0 > 1$) of a generator of $\Gamma(g) = \{\gamma \in \Gamma : g = \pm \gamma g \gamma^{-1}\}$.

---

[2] T. Hiramatsu, On traces of Hecke operators on the space of cusp forms of weight 1. J. Reine Angew. Math. **402** (1989), 166-180.

Bibliography

[1] T. M. Apostol, Modular Functions and Dirichlet Series in Number Theory, Second Ed., Springer, 1990.

[2] T. Arakawa, Selberg zeta functions and the dimensions of the space of elliptic cusp forms of lower weights, Comm. Math. Univ. Sancti Pauli, **39**, 87-109 (1990).

[3] C. Arf, Untersuchungen über quadratische Formen in Körpen der Charakteristik 2, Teil I, J. Reine Angew. Math., **183** (1941), 148-167.

[4] J. Arthur, Automorphic representations and number theory, Canadian Math. Soc., Conference Proceedings, Vol. 1, 3-51 (1981).

[5] M. F. Atiyah, Riemann surfaces and spin structures, Ann. scient. Éc. Norm. Sup., 4e séries, t.4, 47-62 (1971).

[6] T. Barnet-Lamb, D. Geraghty, M. Harris, and R. Taylor, A family of Calabi-Yau varieties and potential automorphy II, Publ. Res. Inst. Math. Sci. **47** (2011), 29-98.

[7] J. Bernstein and S. Gelbart (ed.), An Introduction to the Langlands Program, Birkhäuser, 2003.

[8] J. P. Buhler, Icosahedral Galois Representations, Lecture Notes in Math. **654**, Springer, 1978.

[9] D. Bump, W. Duke, J. Hoffstein, and H. Iwaniec, An estimate for the Hecke eigenvalues of Maass forms, International Math. Research Notices, No. 4, 75-81 (1992).

[10] K. Buzzard, M. Dickinson, N. Shepherd-Barron, and R. Taylor, On icosahedral Artin representations, Duke Math. J., **109**, 283-318 (2001).

[11] W. Casselman, GL_n, Proc. Symposium on Algebraic Number Fields (ed. A. Fröhlich), Academic Press, 1977, 663-704.

[12] T. Chinburg, Stark's conjecture for L-functions with first-order zeros

at $s = 0$. Adv. in Math., **48** (1983), 82-113.

[13] S. Chowla and M. Cowles, On the coefficient c_n in the expansion $x\prod_1^\infty(1-x^n)^2 \times (1-x^{11n})^2 = \sum_1^\infty c_n x^n$, J. Reine Angew. Math., **292** (1977), 209-220.

[14] U. Christian, Über elliptische Sptizenformen von Gewicht 1, Comm. Math. Univ. Sancti Pauli, **38**, 169-196 (1989).

[15] ——, Untersuchung Selbergscher Zetafunktionen, J. Math. Soc. Japan, **41**, 503-537 (1989).

[16] J. W. Cogdell, H. H. Kim, and M. R. Murty, Lectures on Autmorphic L-functions, Fields Institute Monographs, **20**, AMS, 2004.

[17] P. Deligne et J.-P. Serre, Formes modulaires de poids 1, Ann. scient. Éc. Norm. Sup., 4e séries, t.**7** (1974), 507-530.

[18] J.-M. Deshouillers and H. Iwaniec, Kloosterman sums and Fourier coefficients of cusp forms, Invent. Math., **70**, 219-288 (1982).

[19] R. H. Dye, On the Arf invariant, J. Algebra, **53** (1978), 36-39.

[20] I. Y. Efrat, The Selberg trace formula for $\mathrm{PSL}_2(\mathbb{R})^n$, Memoirs of AMS, **65** (1987).

[21] M. Eichler, Der Hilbertsche Klassenkörper eines imaginärquadratisch -en Zahlkörper, Math. Z., **64** (1956). 229-242.

[22] E. Freitag, Hilbert Modular Forms, Springer-Verlag, Berlin Heidelberg, 1990.

[23] G. Frey, Construction and arithmetical applications of modular forms of low weight, Centre de Recherches Math. CRM Proceedings & Lecture Notes, Vol. **4** (1994), AMS, 1-21.

[24] R. Fricke, Lehrbuch der Algebra III, Braunschweig, 1928.

[25] S. Gelbart, Automorphic forms and Artin's conjecture, Lecture Notes in Math., **627**, Springer-Verlag, 1978, 242-276.

[26] —— and H. Jacquet, Forms of GL(2) from the analytic point of view, Proc. of Symposia in Pure Math., **33** (1979), Part I, 213-251.

[27] ——, Lectures on the Arthur-Selberg Trace Formula, Univ. Lecture Series, Vol. **9**, AMS, 1996.

[28] ——, Three lectures on the modularity of $\bar{\rho}_{E,3}$ and the Langlands reciprocity conjecture, Modular Forms and Fermat's Last Theorem (ed. G. Cornell, J. H. Silverman and G. Stevens), Springer-Verlag, New York (1997), 155-207.

[29] B. H. Gross and J. Harris, On some geometric constructions related to theta characteristics, 279-311 in 'Contributions to Automorphic Forms, Geometry and Number Theory' (Ed. H. Hida, D. Ramakrishnan, and F. Shahidi), The Johns Hopkins Univ. Press, 2004.

[30] R. C. Gunning, Lectures on Modular Forms, Annals of Math. Studies, Number **48**, Princeton Univ. Press, 1962.

[31] H. Hasse, Neue Begründung der komplexen Multiplikation, I, J. Reine Angew. Math., **157** (1927), 115-139; II, Ibid., **165** (1931), 64-88.

[32] ——, Uber der Klassenkorper zum quadratischen Zahlkorper mit der Diskriminante -47, Acta Arith., **9** (1964), 419-434.

[33] —— and J. Liang, Uber den Klassenkorper zum quadratischen Zahlkorper mit der Diskriminante −47 (Fortsetzung), Acta Arith., **16** (1969), 89-97.

[34] E. Hecke, Über einen neuen Zusammenhang zwischen elliptischen Modulfunktionen und indefiniten quadratischen Formen, Math. Werke, no. **22** (1925), 418-427.

[35] ——, Theorie der elliptischen Modulfunktionen, Ibid., no. **23** (1926), 428-460.

[36] ——, Über das Verhalten von $\sum_{m,n} e^{2\pi i \tau \frac{|m^2-2n^2|}{8}}$ und ähnlichen Funktionen bei Modulsubstitutionen, Math. Werke, no. **25** (1927), 487-498.

[37] D. A. Hejhal, The Selberg trace formula for PSL(2, $\mathbb{R}$) I, Lecture Notes in Math., **548**, Springer-Verlag, 1976.

[38] T. Hiramatsu, Eichler classes attached to automorphic forms of dimension −1, Osaka J. Math., **3** (1966), 39-48.

[39] —— and Y. Mimura, On automorphic forms of weight one. II. The Arf invariant and d_0 mod 2. Math. Sem. Notes Kobe Univ. **9** (1981), 259-267.

[40] ——, Higher reciprocity laws and modular forms of weight one. Comment. Math. Univ. St. Pauli, **31** (1982), 75-85.

[41] ——, On some dimension formula for automorphic forms of weight one. I, Nagoya Math. J., **85** (1982), 213-221.

[42] —— and N. Ishii, Quartic residuacity and cusp forms of weight one, Comment. Math. Univ. St. Pauli, **34** (1985), 91-103.

[43] —— and Y. Mimura, The modular equation and modular forms of weight one, Nagoya Math. J., **100** (1985), 145-162.

[44] ——, N. Ishii, Y. Mimura, On indefinite modular forms of weight one, J. Math. Soc. Japan, **38** (1986), 67-83.

[45] ——, On some dimension formula for automorphic forms of weight one II, Nagoya Math. J., **105** (1987), 169-186.

[46] —— and S. Akiyama, On some dimension formula for automorphic forms of weight one III, Nagoya Math. J., **111** (1988), 157-163.

[47] ——, Theory of automorphic forms of weight 1, Advanced Studies in Pure Math., **13** (1988), 503-584.

[48] ——, A formula for the dimension of spaces of cusp forms of weight 1, Advanced Studies in Pure Math., **15** (1989), 287-300.

[49] ——, On traces of Hecke operators on the space of cusp forms of weight 1, Crelles **402** (1989), 166-180.

[50] N. Ishii, Cusp forms of weight one, quartic reciprocity and elliptic curves, Nagoya Math. J., **98** (1985), 117-137.

[51] H. Ishikawa, On the trace formula for Hecke operators, J. Fac. Sci. Unv. Tokyo, Sec. IA, **20** (1973), 217-238.

[52] ——, On the trace of Hecke operators for discontinuous groups operating on the product of the upper half planes, J. Fac. Sci. Univ. Tokyo, Sec. IA, **21** (1974), 357-376.

[53] ——, The traces of Hecke operators in the space of the 'Hilbert modular' type cusp forms of weight two, Sci. Papers Coll. Gen. Ed. Univ. Tokyo, **29** (1979), 1-28.

[54] H. Iwaniec, Small eigenvalues of Laplacian for $\Gamma_0(N)$, Acta Arithmetica, Vol. **LVI** (1990), 65-82.

[55] H. Jacquet and R. P. Langlands, Automorphic forms on GL(2), Lecture Notes in Math., **114**, Springer-Verlag. 1970.

[56] M. Jimbo and T. Miwa, Irreducible decomposition of fundamental modules for $A_i^{(1)}$ and $C_l^{(1)}$ and Hecke modular forms, Publ. Res. Inst. Math. Sci., **434**, Kyoto Univ., 1983.

[57] D. Johnson, Spin structures and quadratic form on surfaces, J. London Math. Soc., **22** (1980), 365-373.

[58] V. G. Kac and D. H. Peterson, Infinite-dimensional Lie algebras, theta functions and modular forms, Adv. in Math., **53** (1984), 125-254.

[59] M. Kisin, Modularity of 2-dimensional Galois representations, Current Developments in Math., vol. 2005 (2007), 191-230.

[60] A. W. Knapp, Introduction to the Langlands program, Proc. of Symposia in Pure Math., Vol. **61** (1997), 245-302.

[61] F. H. Koch, Arithmetische Theorie der Normkörper von 2-Potenzgrad mit Diedergruppe, J. Number Theory, **3** (1971), 412-443.

[62] M. Koike, Higher reciprocity law, modular forms of weight 1 and elliptic curves, Nagoya Math. J., **98** (1985), 109-115.

[63] T. Kubota, Elementary Theory of Eisenstein Series, Kodansha and Halsted, Tokyo-New York, 1973.

[64] R. P. Langlands, Base Change for GL(2), Ann. of Math. Studies, no. **96**, Princeton Univ. Press, Princeton, 1980.

[65] G. Lion and M. Vergne, The Weil representation, Maslov index and theta series, Boston: Birkhäuser, 1980.

[66] H. Maass, Über die neue Art von nichtanalytischen automorphen Functionen und die Bestimmung Dirichlet Reihen durch Funktionalgleichungen, Math. Annalen, Bd. **121**, 141-183 (1949).

[67] ——, Lectures on Modular Functions of One Complex Variable, Tata Institute, Springer 1983.

[68] C. J. Moreno, The higher reciprocity laws: an example, J. Number Theory, **12** (1980), 57-70.

[69] ——, The value of $L(1/2, \chi)$ for Abelian L-functions of complex quadratic fields, J. Number Theory, **18** (1984), 269-288.

[70] ——, Advanced Analytic Number Theory: L-functions, Math. Surveys and Monographs, Vol. **115** (2005), AMS.

[71] D. Mumford, Theta characteristics of an algebraic curve, Ann. scient. Éc. Norm. Sup., 4e séries, t. **4** (1971), 181-192.

[72] V. K. Murty, Lacunarity of modular forms, J. of the Indian Math. Soc., **52**, 127-146 (1987).

[73] A. Ogg, Modular Forms and Dirichlet Series, Benjamin, 1969.

[74] V. Pasol and A. Polishchuk, Universal triple Massey products on elliptic curves and Hecke's indefinite theta series, Mosc. Math. J., **5** (2) (2005), 443-461.

[75] H. Petersson, Über Eisensteinsche Reihen und automorphe Formen von der Deimension -1, Comment. Math. Helv., **31** (1956), 111-144.

[76] A. Polishchuk, A new look at Hecke's indefinite theta series, Contemporary Math., **291**, 183-191 (2001).

[77] ——, Indefinite theta series of signature $(1,1)$ from the point of view of homological mirror symmetry, Adv. Math., **196** (1) (2005), 1-51.

[78] J. D. Rogawski, Functoriality and the Artin conjecture. Proc. of Symposia in Pure Math., Vol. **61** (1997), 331-353.

[79] l. J. Rogers, Second memoir on the expansion of certain infinite products. Proc. London Math. Soc., **25** (1894), 318-343.

[80] P. Sarnak, Additive number theory and Maass forms, Lecture Notes in Math., **1052** (1984), Springer-Verlag, 286-309.

[81] ——, Selberg's eigenvalue conjecture, Notices of the AMS, Vol. **42**, Number **11** (1995), 1272-1297.

[82] ——, Maass cusp forms with integer coefficients, 121-127 in 'A Panorama of Number Theory' (Ed. Wustholz), Cambridge, 2002.

[83] ——, Spectra of hyperbolic surfaces, Bulletin of the AMS, **40**, 441-478 (2003).

[84] M. Sato, Theory of hyperfunctions, I, J. Fac. Sci. Univ. Tokyo, Sec. I, **8** (1959), 139-193.
[85] A. Selberg, Harmonic analysis, Göttingen Lecture Notes (1954), Collected papers of A. Selberg vol.I, 626–674, Springer-Verlag, Berlin, 1989.
[86] ——, Harmonic analysis and discontinuous groups in weakly symmetric Riemannian spaces with applications to Dirichlet series, J. Indian Math. Soc., **20** (1956), 47-87.
[87] ——, Discontinuous groups and harmonic analysis, Proc. Int. Math. Congr. Stockholm, 1962, 177-189.
[88] ——, On the estimation of Fourier coefficients of modular forms, Proc. of Symposia in Pure Math., **8** (1965), 1-15.
[89] J.-P. Serre, Modular forms of weight one and Galois representations, Proc. Symposium on Algebraic Number Fields (ed. A. Fröhlich), Academic Press, 1977, 193-268.
[90] ——, Sur le représentations modulaires de degré 2 de $\mathrm{Gal}(\overline{\mathbb{Q}}/\mathbb{Q})$, Duke Math. J., **54**(1), 179–230 (1987).
[91] ——, Abelian ℓ-Adic Representations and Elliptic Curves, Addison Wesley, 1989.
[92] F. Shahidi, Symmetric power L-functions for GL(2), Centre de Recherches Math. CRM Proceedings & Lecture Notes, Vol. **4** (1994), 159-182.
[93] ——, Functoriality and small eigenvalues of Laplacian on Riemann surfaces, Survey in Differential Geometry, **IX** (2004), International Press, 385-400.
[94] H. Shimizu, On discontinuous groups operating on the product of the upper half planes, Annals of Math., **77** (1963), 33-71.
[95] ——, On traces of Hecke operators, J. Fac. Sci. Univ. Tokyo, Sec. I, **10** (1963), 1-19.
[96] ——, A remark on the Hilbert modular forms of weight 1, Math. Ann., **265** (1983), 457-472.
[97] T. Shintani, On certain ray class invariants of real quadratic fields, J. Math. Soc. Japan, **30** (1978), 139-167.
[98] H. M .Stark, Class fields and modular forms of weight one, Lecture Notes in Math., **601**, Springer-Verlag, 1977, 277-287.
[99] ——, On modular forms of weight one from real quadratic fields and theta functions, J. Ramanujan Math. Soc., **3**(1), 63–79 (1988), Ramanujan Birth Centenary Special Issue.
[100] F. Strömberg, Maass waveforms on $(\Gamma_0(N), \chi)$ (Computational As-

pect), Ch. 6 in Hyperbolic Geometry and Applications in Quantum Chaos and Cosmology, LMS Lecture Notes Series, No. **397**, 2011.
[101] L. A. Takhtajan and A. I. Vinogradov, The Gauss-Hasse hypothesis on real quadratic fields with class number one, Crelles **335** (1982), 40-86.
[102] R. Taylor, On icosahedral Artin representations II, Amer. J. of Math., **125**(3), 549–566 (2003).
[103] Y. Tanigawa, On cusp forms of octahedral type, Proc. Japan Akad., **57**, Ser. A, no. **7** (1986), 270-273.
[104] —— and H. Ishikawa, The dimension formula of the space of cusp forms of weight one for $\Gamma_0(p)$, Nagoya Math. J., **111** (1988), 115-129.
[105] T. Tunnell, Artin's conjecture for representations of octahedral type, Bull. Amer. Math. Soc., **5** (1981), 173-175.
[106] ——, A classical Diophantine problem and modular forms of weight 3/2, Invent. Math., **72** (1983), 323-334.
[107] M.-F. Vignéra, Représentations Galoisiennes paires, Glasgow Math. J., **27** (1985), 223-237.
[108] H. Weber, Lehrbuch der Algebra III, Braunschweig, 1908.
[109] A. Weil, Über die Bestimmung Dirichletscher Reihen durch Funktionalgleichungen, Math. Annalen, **168** (1967), 149-156.
[110] B. F. Wyman, What is a reciprocity law? Amer. Math. Monthly, **79** (1972), 571-586.
[111] H. Zassenhaus and J. Liang, On a problem of Hasse, Math. comp., **23** (1969), 515-519.
[112] P. Zograf, Selberg trace formula for the Hilbert modular group of a real quadratic algebraic number field, J. of Soviet Math., **19** (1982), 1637-1652.
[113] Seminar on Complex Multiplication, Lecture Notes in Math., **21**, Springer-Verlag, 1966 (for the Institute Seminar by Borel, Chowla, Herz, Iwasawa and Serre, held in 1957-1958).

[illegible] (2005), Chapter [illegible] Geometry and Applications of [illegible] Chaos [illegible] LMS Lecture Notes [illegible]
[101] [illegible] A. Tak[illegible] and [illegible] Van[illegible], The [illegible] on re[illegible] [illegible] (1993) [illegible]
[102] F. [illegible] A [illegible] representation [illegible] 125, [illegible] (2008) [illegible]
[103] X. [illegible] [illegible] 37 [illegible] (2011) 370 [illegible]
[104] [illegible]
[105] [illegible] 1998 [illegible]
[106] [illegible] 25 [illegible] (2012) [illegible]
[107] [illegible] (1998) [illegible]
[108] [illegible]
[109] [illegible] (1995) [illegible] [illegible]
[110] [illegible] 125 [illegible]
[111] F. Zo[illegible] [illegible] [illegible] (1997) 1637–1657.
[112] [illegible] Lecture Notes in Math. [illegible] [illegible] Lond. Phys. [illegible]

Index

Printed in the United States
By Bookmasters